IMAGES
of America

FORK SHOALS

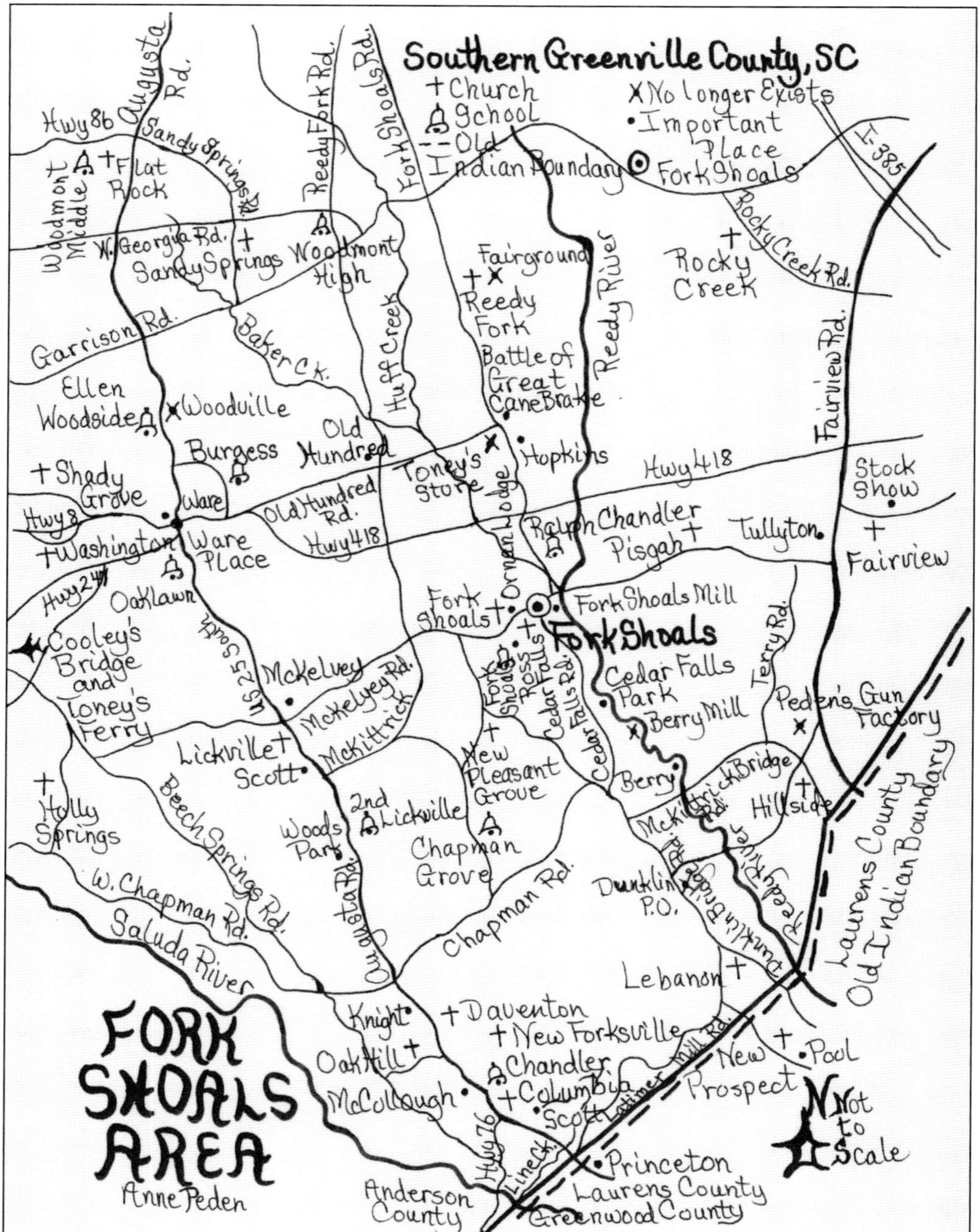

This map of the Fork Shoals area is courtesy of Anne Peden.

On the Cover: Today, Fork Shoals School is the oldest school in the southern part of Greenville County founded in the 1800s. It began as a one-room school on the grounds of Fork Shoals Baptist Church. This photograph shows the new school building that was completed in 1916. The building was remodeled and eventually replaced in 1998. (Courtesy of Fork Shoals Elementary School.)

IMAGES
of America

Fork Shoals

Fork Shoals Historical Society

ARCADIA
PUBLISHING

ISBN 978-0-7385-9083-7

Published by Arcadia Publishing
Charleston, South Carolina

Printed in the United States of America

Library of Congress Control Number: 2012943712

For all general information, please contact Arcadia Publishing:
Telephone 843-853-2070
Fax 843-853-0044
E-mail sales@arcadiapublishing.com
For customer service and orders:
Toll-Free 1-888-313-2665

Visit us on the Internet at www.arcadiapublishing.com

Fork Shoals Historical Society dedicates this book to those who have gone before us who made this history, to the current communities, and to the children who will be the future of the Fork Shoals area.

Contents

ACKNOWLEDGMENTS

The Fork Shoals area is comprised of many caring souls and wonderful folks who have shared their lives and stories with us over the past two years. The Fork Shoals Historical Society is privileged to have been given the trust of so many local citizens. The photographs in this pictorial history are each credited with a courtesy line at the end of the caption, so for those folks, we will let that suffice as recognition for their generosity. Knowing that many people provided photographs that are not in this book, we regret that more could not be included. Many hours were spent looking through photographs, grouping them, scanning them, checking them with our editor at Arcadia, writing captions, and generally "oohing" and "aahing" over them. We have learned so many stories and made so many connections during this process that cannot be covered herein. A true effort was made to select the best photographic history from submissions donated. We hope that our collection of histories, photographs, and artifacts continues to grow.

There are people who, as part of the book committee, have worked for months to aid in the preparation of this document. They include Tuffy Atkins, Carol Gilley, Debbie and Ed Moorer, Anne Peden, Kathy Saxon, Jim Scott, and Woody Woods. Along with those who worked from the beginning, Virginia Dean and Cheryl MacKnight have also spent countless hours in order to make this the best book possible. Cheryl MacKnight provided the Fork Shoals School historical collections of her mother, Peggy Sims Smith. In addition, Edwin Terry, Katherine Ross, Gloria Hughey, Kim Reid, Charles Heatherly, and Edward Ridgeway all have been extremely helpful in many ways.

Without the aid of Erik Bianchi of Bianchi Video Productions, the prompt completion of this history would have been impossible. Maggie Bullwinkel, our contact at Arcadia, has been gracious and understanding as she guided us through this process. Thank you, both, for all the technical aid.

Finally, the committee members want to thank our family members for their understanding. Thanks go to all you wonderful folks.

INTRODUCTION

Prior to 1776, the areas surrounding the Reedy River were used as Cherokee hunting grounds. Native Americans used the rock shoals of the Reedy River at the current Cedar Falls Park as a hunting campsite and a ford to cross the river. Many artifacts from this era have been found along the banks of the Reedy River and throughout the surrounding countryside.

On December 22, 1775, the Battle of the Great Cane Brake was fought just north along the river. This was the only Revolutionary War battle in Greenville County. Often called the "Snow Campaign" because a foot of snow complicated the battle, the Patriots routed the Loyalist forces while suffering only one wounded soldier themselves. A Daughters of the American Revolution plaque now designates this site just north of Highway 418 on Fork Shoals Road.

Settlers and traders had already been present in the Indian Territory prior to the Revolutionary War, but in 1777, the Treaty of DeWitt's Corner secured most of the remaining portion of Cherokee lands in the upstate for the colony. Europeans, who were given large land grants by the new government as payment for service during the war, established in or near Fork Shoals the first settlements in Greenville County in the 1780s. Many of these soldiers founded the earliest churches, beginning in 1780 with Fork Shoals Baptist Church, followed by Lebanon United Methodist in 1785, and Fairview Presbyterian, the first Presbyterian church in Greenville County, in 1786.

There were two cotton mills along Reedy River at the shoals. One was near the Cedar Falls dam. The other, which is currently a private residence, is still standing two miles upstream on McKelvey Road.

Between 1820 and 1852, Hudson Berry and his sons operated a mill at Cedar Falls. It was a small operation, running only 72 spindles. There was also a general store, water-powered gristmill, sawmill, and cotton gin on the premises. Today, you can see the concrete piers that were used for the cotton gin and sawmill.

In 1870, Hewlett Sullivan and his brother Dr. James Sullivan built a mill on what is now McKelvey Road. It was the first new cotton mill in Greenville County after the end of the Civil War (1865). At first, it ran 2,000 spindles. By 1952, after many ownership changes, it was running 14,000 spindles. This mill is often referred to as the Fork Shoals Mill.

However, Fork Shoals Mill was isolated from other local cotton-related industries and was quite a distance from a railway. Therefore, the spinning mill never achieved the level of economic significance of the larger mills. As cotton farming waned, locals traveled to Greenville, Ware Shoals, Fountain Inn, Simpsonville, and Pelzer to work. This smaller mill eventually closed its doors, ceasing production totally in 1978. At that time, it was owned by Riegel Textiles and put 200 employees out of work.

Fork Shoals had not been a typical mill village where residents earned their living solely at the mill. The community had been primarily agrarian, and the people had supplemented their farm income by working at the mill. Many still treasure their farming roots by keeping cattle, horses, and growing gardens. Few large crop farmers still practice their trade, but there are remnants dotting the countryside.

Fork Shoals Baptist Church was first organized in 1777–1780 "in a brush arbor where the Reedy River and Big Creek came together at the shoals which was Fork Shoals." With the church as the foundation of the village, settlers established large plantations cultivating cotton, wheat, corn, oats, livestock, and even silk.

Trading posts were common in the backcountry, and William Toney ran a store (1816–1829) on property that he later sold to John Hopkins in 1834. Along with a store, Hudson Berry built a cotton gin, a gristmill, and a sawmill and harnessed the waters of Cedar Falls to power them. He industrialized to the next level with a small spinning mill to capitalize on the area's primary cash crop, cotton.

Berry's plantation eventually grew to more than 2,000 acres, and with his other businesses, his prosperity was evident. Others benefiting from those cotton-related projects included farmers, farmhands, cotton gin managers, mill operators, and employees in the various businesses as well as in local stores. As companies profited, the village grew and prospered. The citizens were able to build churches, a Masonic Lodge, and a school.

In 1849, the Ornan Lodge No. 69, a fraternal organization of Free Masons, was established in Fork Shoals. The lodge moved several times, originally meeting on the Berry plantation, once being on the grounds of the Fork Shoals Baptist Church (1888), and then, finally, to its current location on McKelvey Road (1983). Two of its members became governors of South Carolina: Robert Cooper (1920–1922) and Carroll Campbell (1986–1994).

Fork Shoals School was established in 1877 on the grounds of the current Fork Shoals Baptist Church. It is believed to have been a one-room log cabin. It moved across the road to its current location in 1916. By 1922, Fork Shoals School, a private high school at this time, became a public school. It is believed to have been the second rural high school in the state to be accredited. Fork Shoals High School merged with nearby Ellen Woodside High School at the Ellen Woodside campus on Highway 25. Fork Shoals Elementary School is now an International Baccalaureate School in a wonderful facility, which opened in 1998.

The freshet (flood) of 1908 had a tremendous, long-lasting impact on the area. It damaged or destroyed most of the buildings along the river—many of which were never rebuilt.

The Fairview Stock Show was a fine example of the agrarian lifestyle in the southern part of Greenville County. Farmers and families gathered to display their best livestock, garden produce, and canned goods; race their fastest horses; and show off their finest attire. It rivaled anything at Tryon or Aiken.

The textile baseball team was an important part of life in Fork Shoals. It brought together the spirit of family, pride, and sense of community from the early 1900s to the mid-1950s. Textile sports were and still are among the strongest ties that bind many former employees to a mill hill.

The annual tractor show brings out the best of the old farming equipment to Fork Shoals School grounds each spring. This festival sponsored by the Fork Shoals Preservation Society began as a fiddling fest, which existed at least as far back as the 1930s and has evolved into the equipment show.

Over the years, the Fork Shoals village was a self-sustaining community and home to various businesses. In the early 20th century, it was home to a mercantile store, bank, post office, about five grocery stores, a washerette, barbershop, hardware store, and filling station. Always at its foundation, there were welcoming places of worship—as there are still today.

Fork Shoals is beginning a new era in its history. Residents, old and new, are investing in the beauty and friendliness of this rural community. Small family businesses are making a living and moving forward. Of course, the talk of the town is Cedar Falls Park, the last and hidden jewel of the parks along the Reedy River in Greenville County. Folks are saying, "Have you seen the new park off McKelvey?"

One

Remarkable Beginnings

Early history

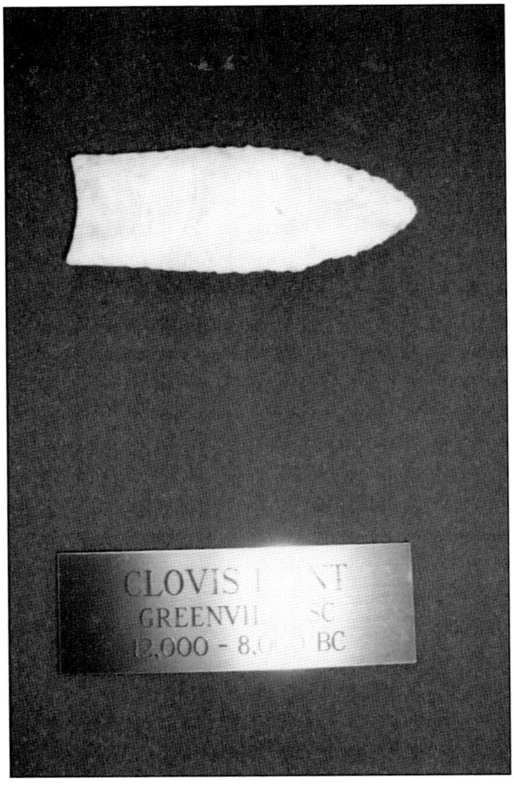

Archaeological evidence suggests that native cultures hunted in the Fork Shoals area at least 10,000 to 12,000 years ago. The oldest man-made artifact, a Clovis point, was unearthed on the Peden farm off McKelvey Road. Clovis points from the Paleo-Indian period date from about 13,000 to 8,000 BC, when the climate was equivalent to that of Southern Canada today. (Courtesy of Richard Sawyer.)

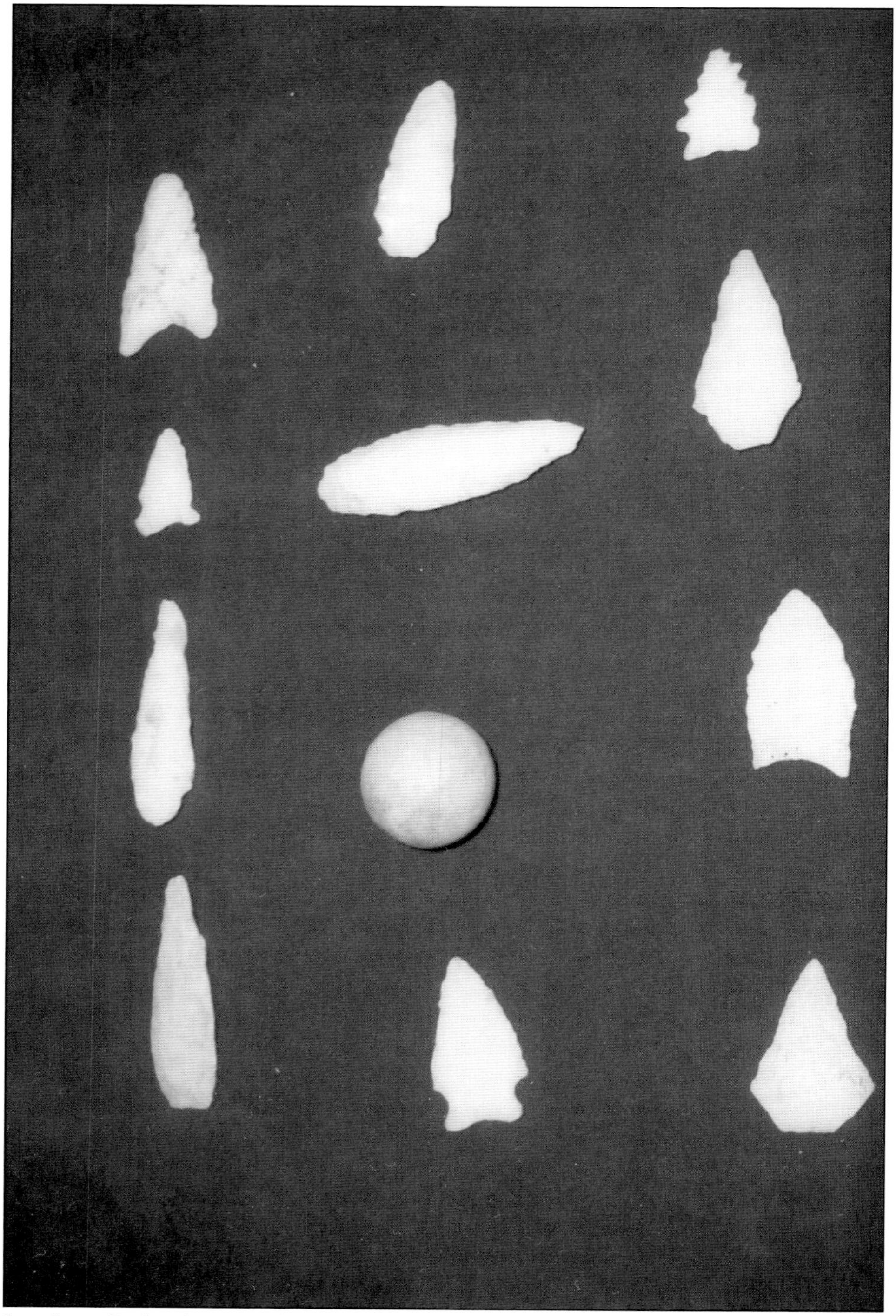

Archaeologist Richard Sawyer's historical research shows that true arrowheads were crafted during the Woodland Period, from 1,000 BC to 1,000 AD. Tribes also constructed bows and arrows. At this time, the Cherokee Indians may have moved into the area. During the Mississippian Period, 1,000 AD to 1,600 AD, more organized groups formed, and the first contact was made with the Europeans. (Courtesy of Michael Ray.)

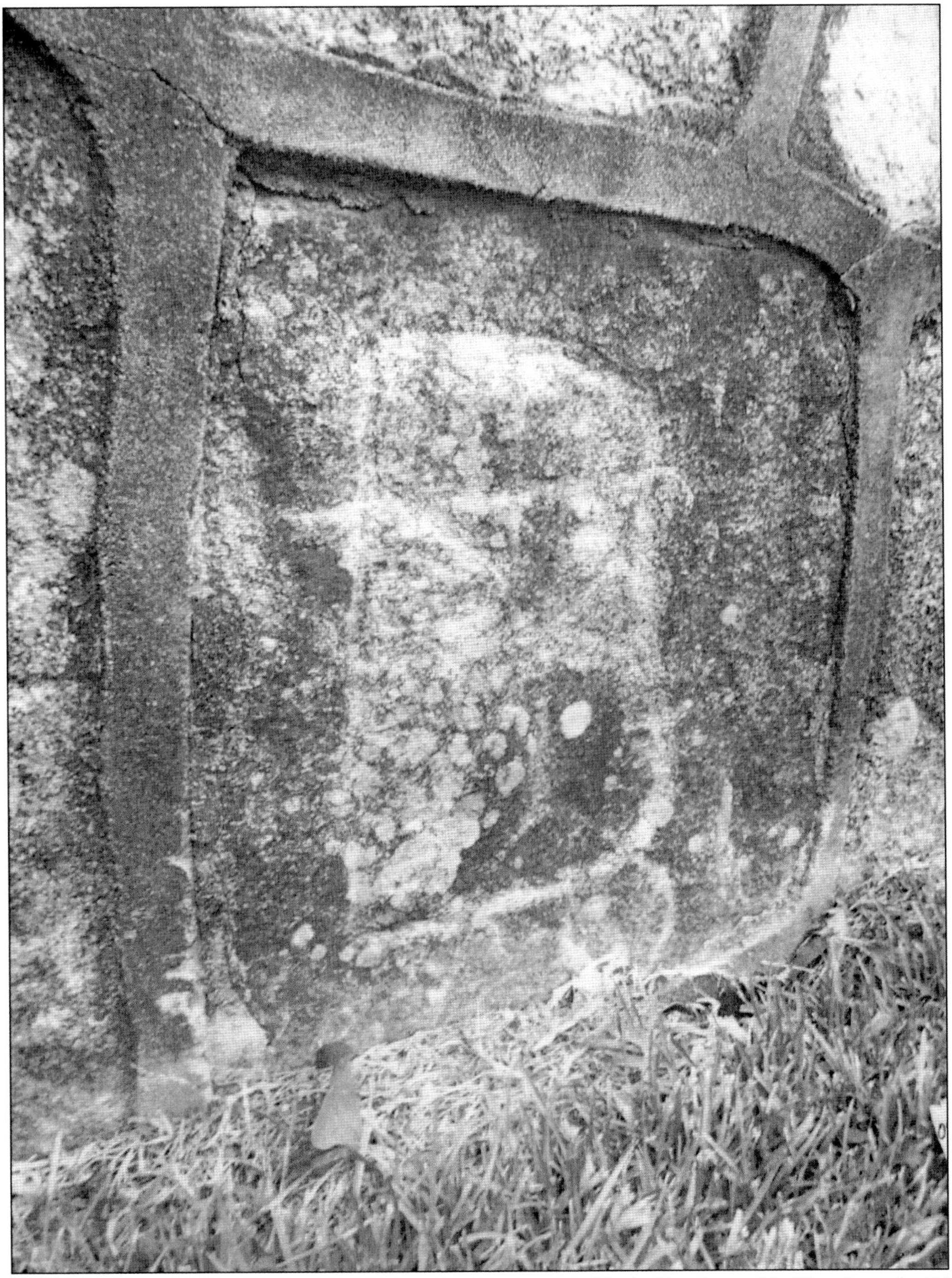

Early settlers in the Fairview Church community just east of Fork Shoals used field rocks to build a fence around the cemetery adjacent to the church. Several of these rocks show petroglyphs, another evidence of early Native American presence in the area. (Courtesy of Rick Owens.)

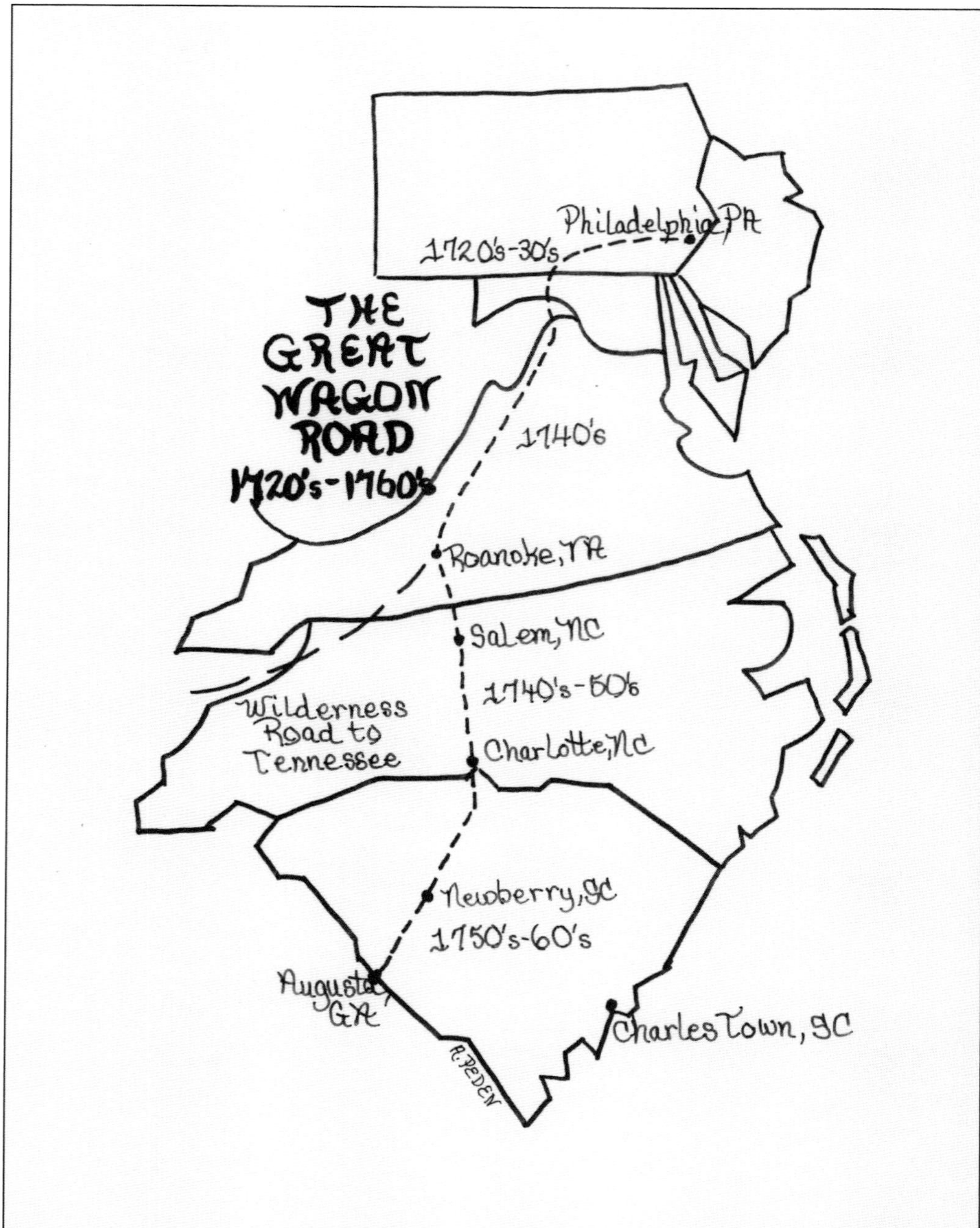

After Europeans began colonizing the Americas, indigenous people lost their lands in South Carolina. This period brought colonists down the Great Wagon Road from Virginia and up the Indian Trails from Charleston into the upstate. Many settled just below Indian Territory in Laurens County, but a few brave souls settled across the boundary prior to the Revolutionary War and the Treaty of DeWitt's Corner. (Courtesy of Anne Peden.)

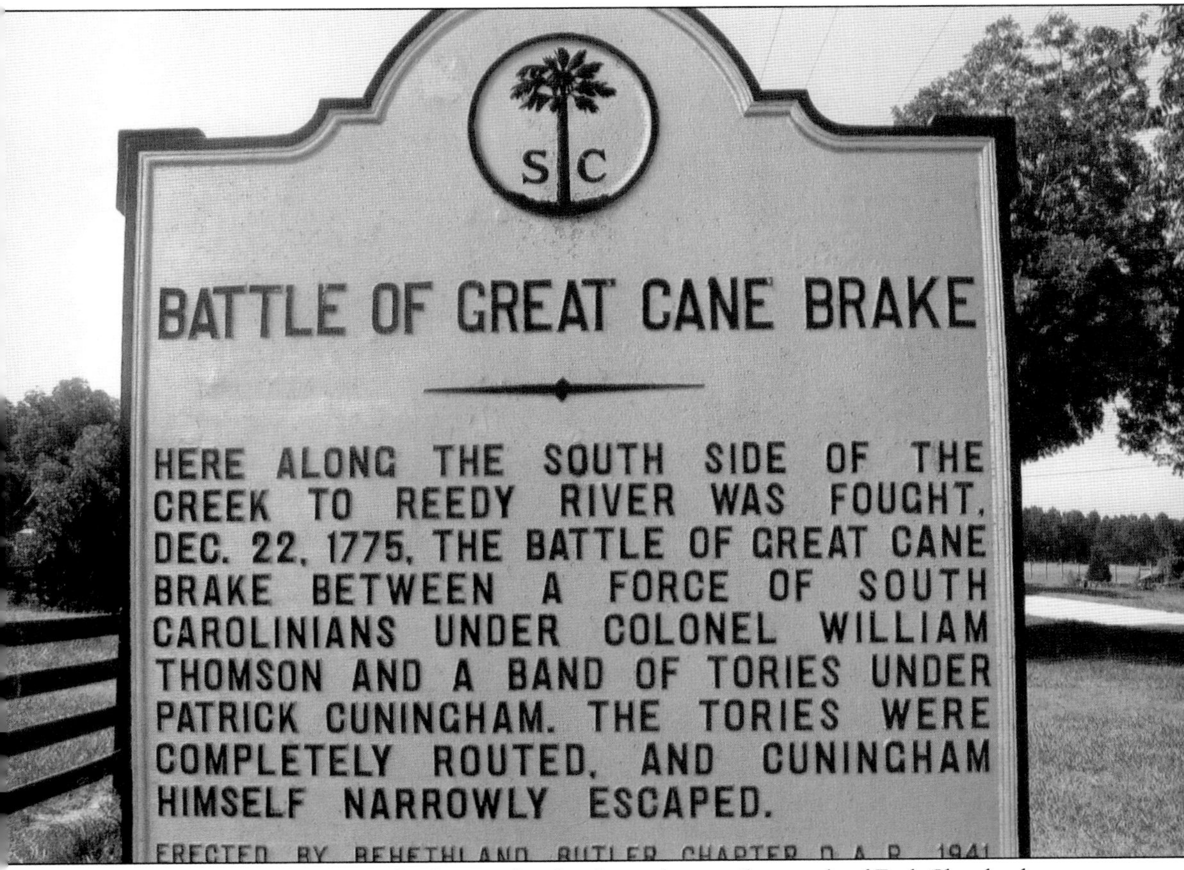

The Battle of the Great Cane Brake was fought about three miles north of Fork Shoals along a creek that feeds into the Reedy River. Often called the Snow Campaign, soldiers fought in almost two feet of snow just prior to Christmas in 1775. After the Revolution, many of them were given land in the old Indian Territory as payment for services rendered. (Courtesy of John Drayton Hopkins's Family.)

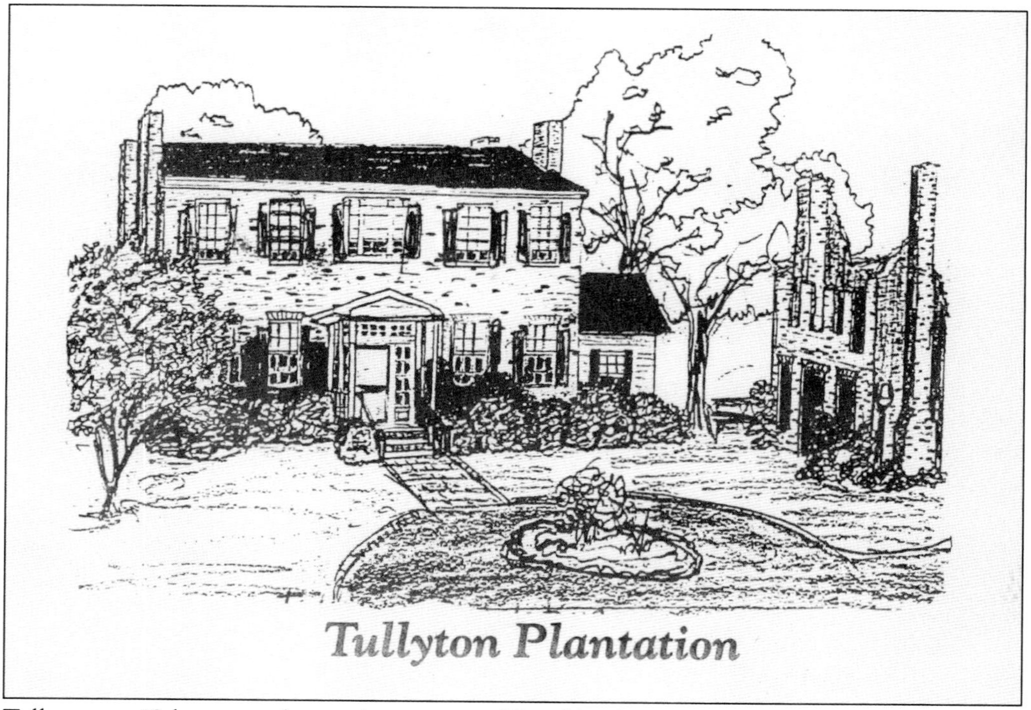

Tullyton, an 18th-century home place, is one of the oldest homes in the area still standing. Ruins of the original building, erected in 1790, remain at the right of the current home, which was built in 1840 for Thaddeus and Louisa Ware Bolling. The original home was used for a post office, tailor shop, insurance office, and kitchen for the big house. (Courtesy of Gale Richardson.)

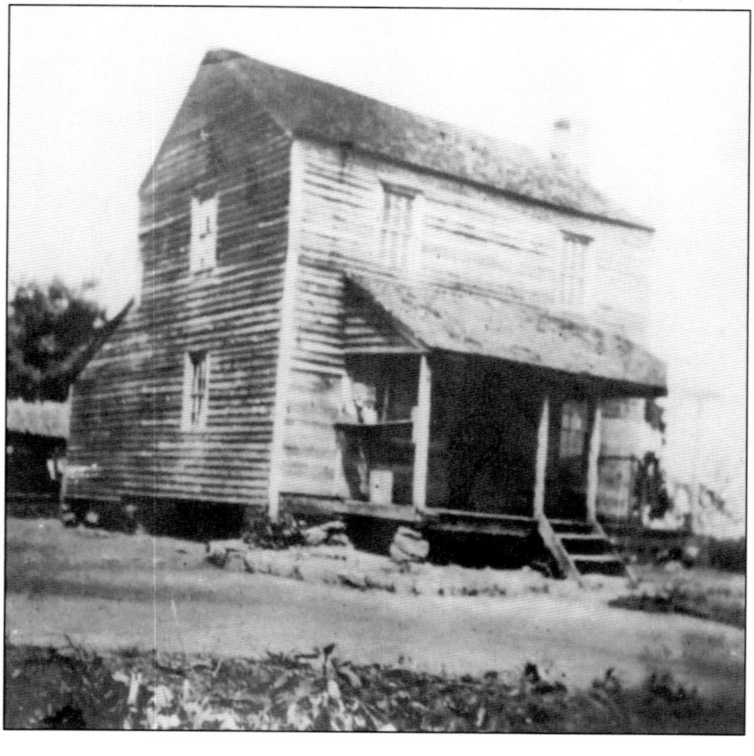

Toney's Store was a two-story log house and an upcountry Indian Trading Post. William Toney sold the store and land to the Hopkins family, where a marker is situated along the road today. In his journal, Vardry McBee, a prominent citizen of Greenville, stated, "Toney was the most prosperous merchant here by far." (Courtesy of John Drayton Hopkins's Family.)

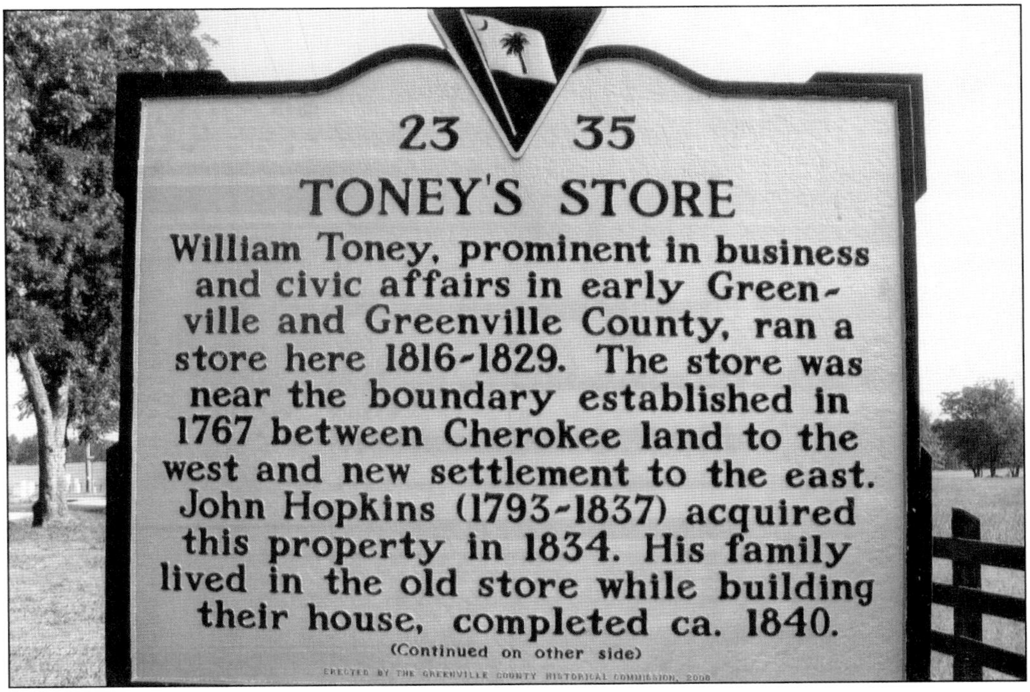

These historical markers line Fork Shoals Road just north of Highway 418. (Both, courtesy of John Drayton Hopkins's Family.)

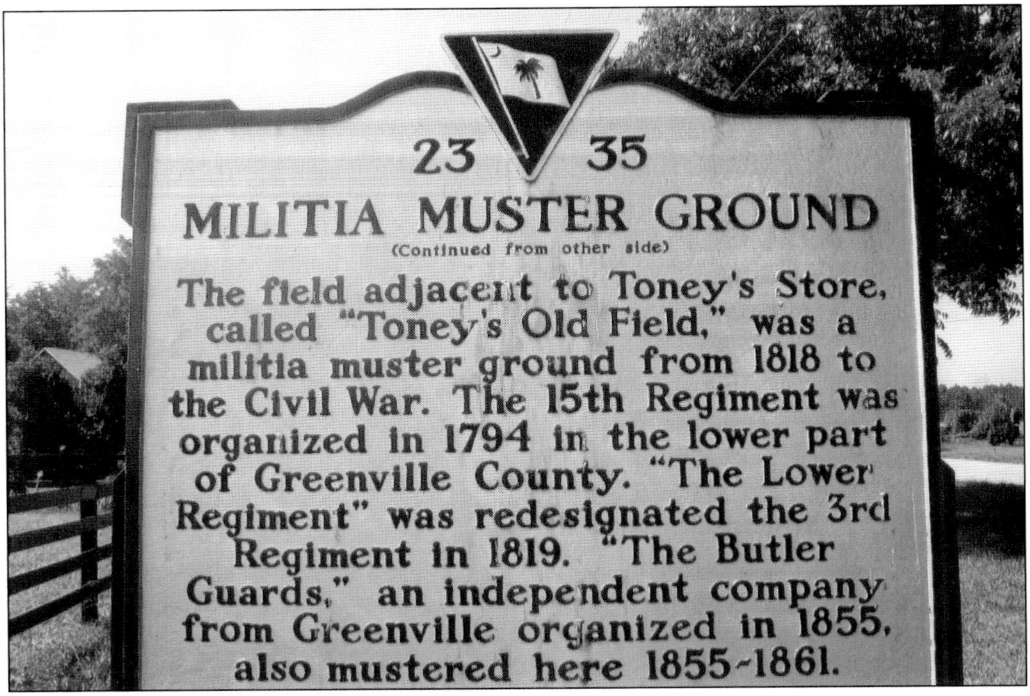

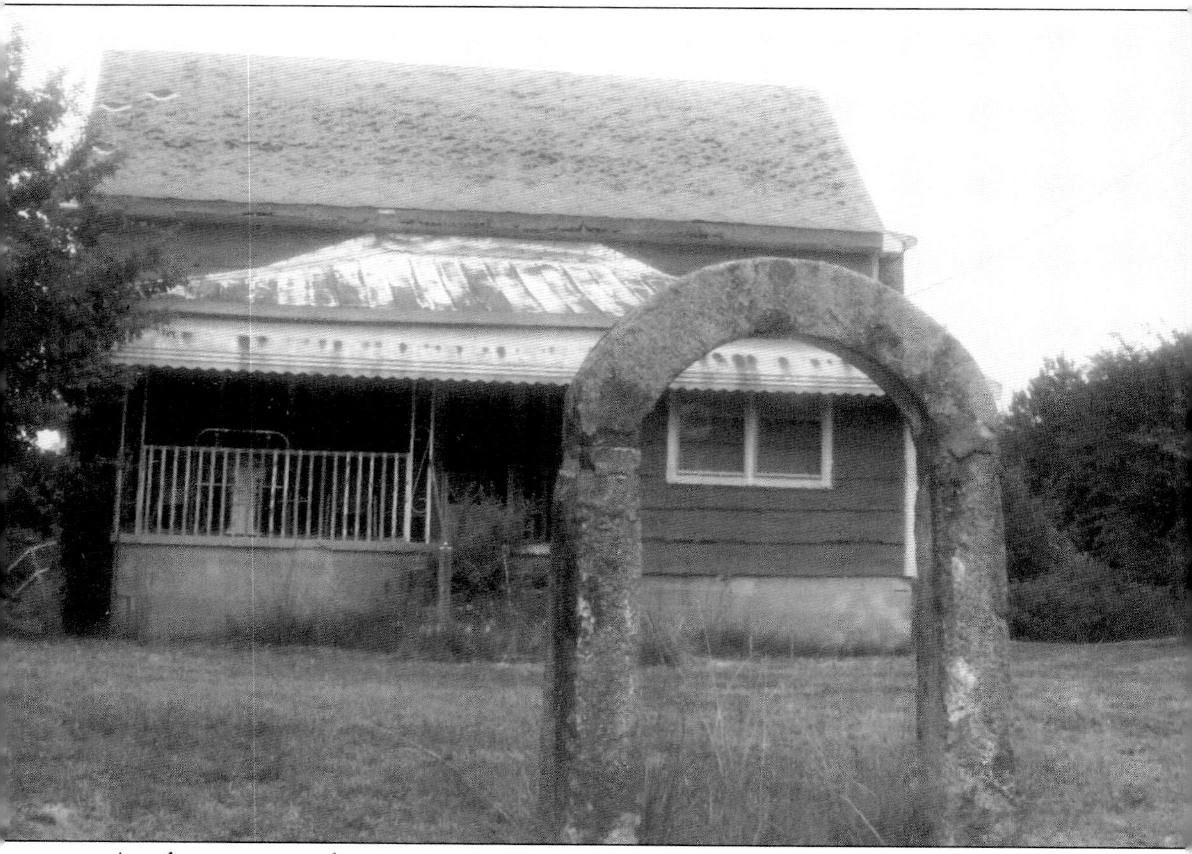

Another prominent businessman gravitated into Fork Shoals from Laurens County and set up several businesses along the Reedy River at Cedar Falls Park. Hudson Berry's home place is located south of the falls, and the concrete arch marks the original location. The home that is currently behind the arch has been added on to several times and does not resemble Berry's plantation house. (Courtesy of Angela Cox Photography.)

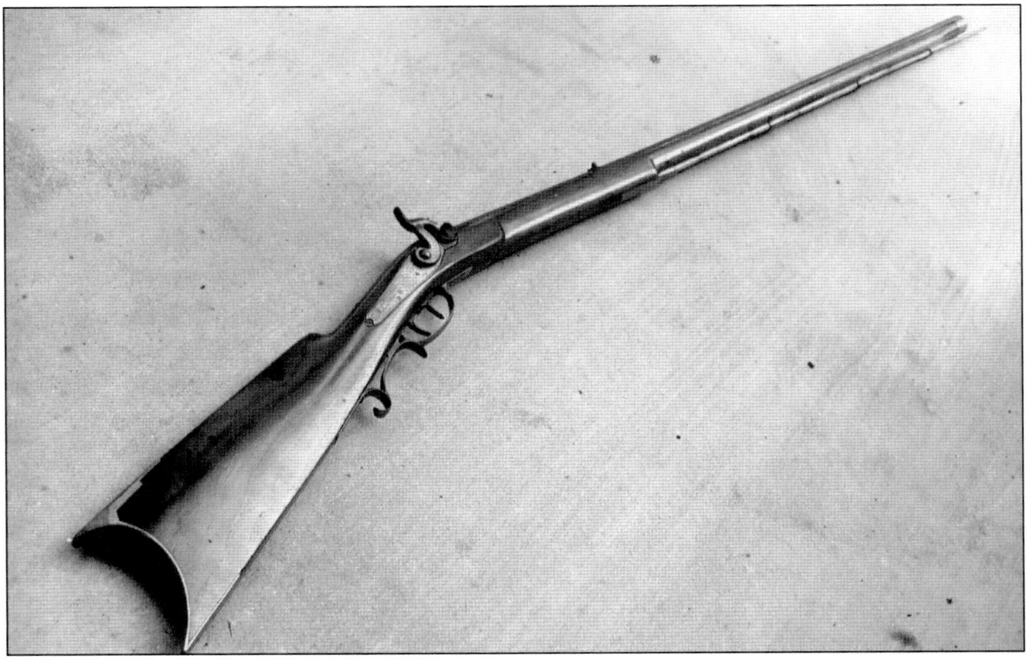

This is one of the remaining Peden Rifles made around 1820. Thomas Peden was its designer. The 40-caliber musket rifle barrel was formed in the creek bed along Terry Road. The stock is of cherry. Peden's Gun Factory was designated on the Geognostic Map of Greenville District, South Carolina, in 1858. (Courtesy of H. and L. Peden with photograph by Carol Gilley.)

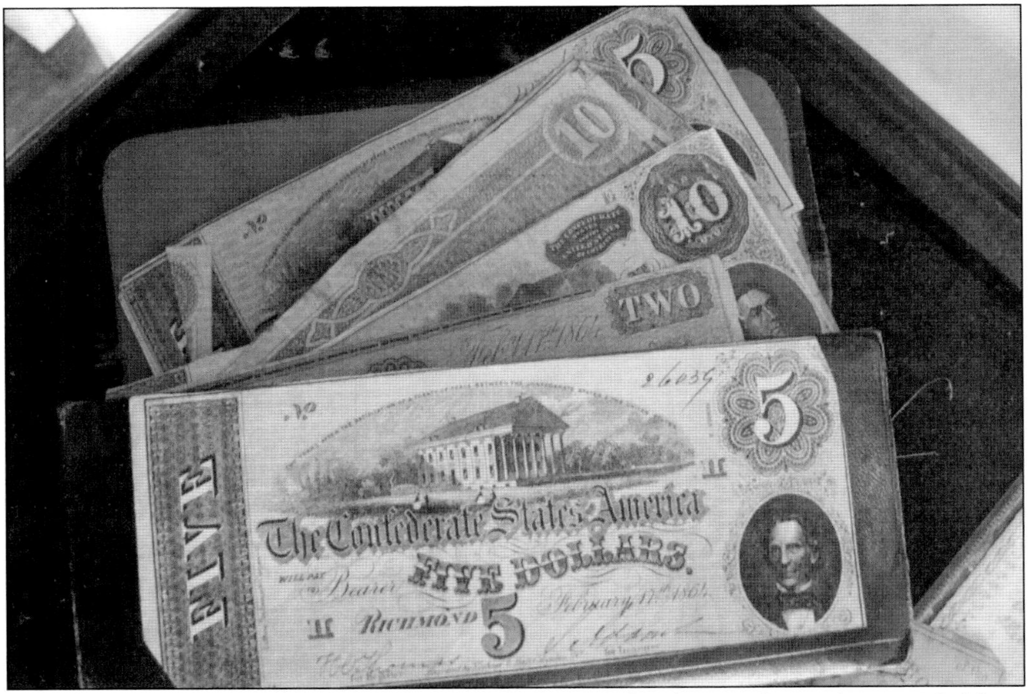

The Confederate States of America issued Confederate bills after South Carolina seceded from the Union and the Confederacy was formed. (Courtesy of H. and L. Peden with photograph by Carol Gilley.)

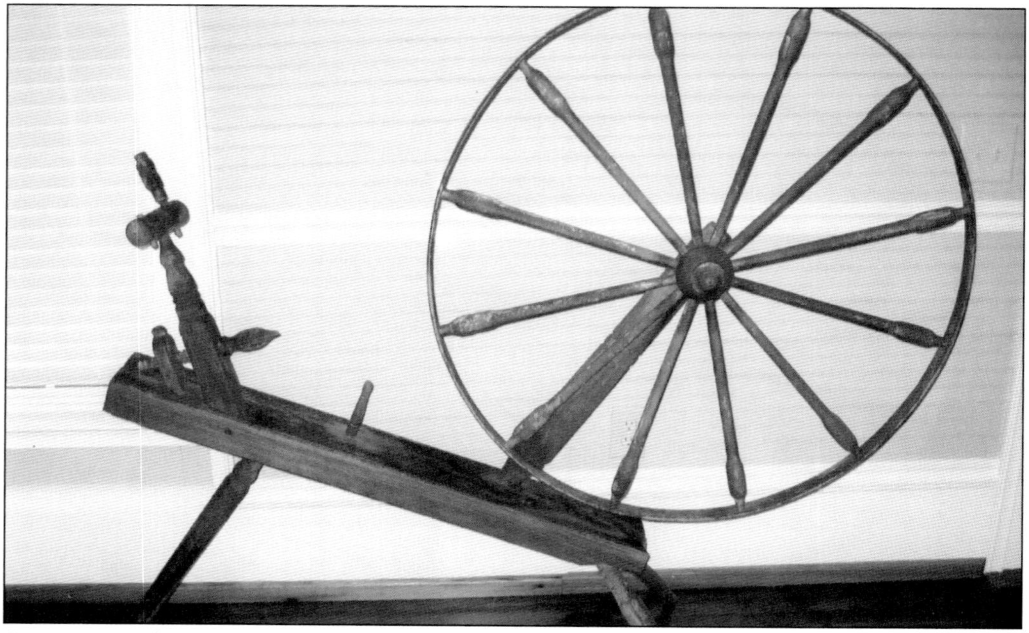

Spinning wool, cotton, linen, and even silk, into yarn was part of the work of the women of the house in early traditions. Spinning wheels were a necessary part of farm life. (Courtesy of H. and L. Peden with photograph by Carol Gilley.)

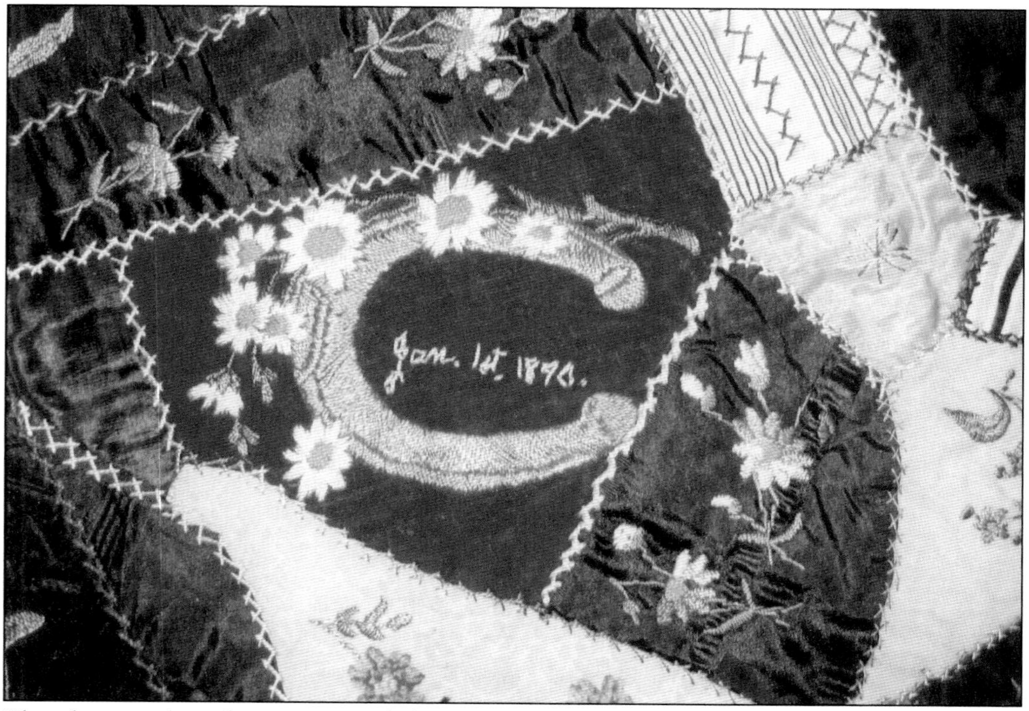

This photograph is of a marriage quilt (a crazy quilt style) that was begun on January 1, 1890, and finished on June 30, 1890. Portions of it were from various dresses and cloth. Some fabric was from a wedding dress. (Courtesy of the H. and L. Peden with photograph by Carol Gilley.)

Two

Guiding Lights

Churches

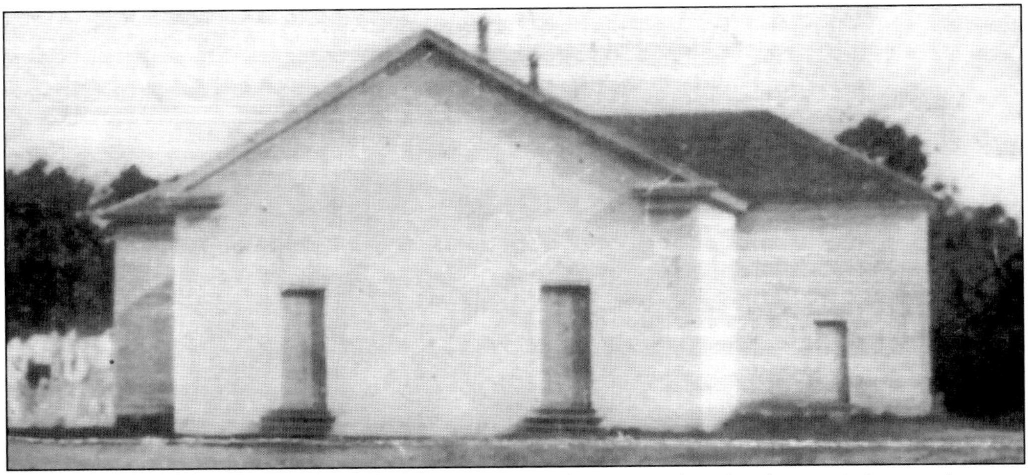

Fork Shoals Baptist Church, established 1777–1780, began as a brush arbor with logs and stones for seating. The name became used to describe the area near the shoals where Big Creek and the Reedy River converge. The second chapel was a log house. A frame building was constructed in 1836. In 1860, a new house of worship was completed; Micajah Berry gave the land. (Courtesy of Fork Shoals Baptist Church.)

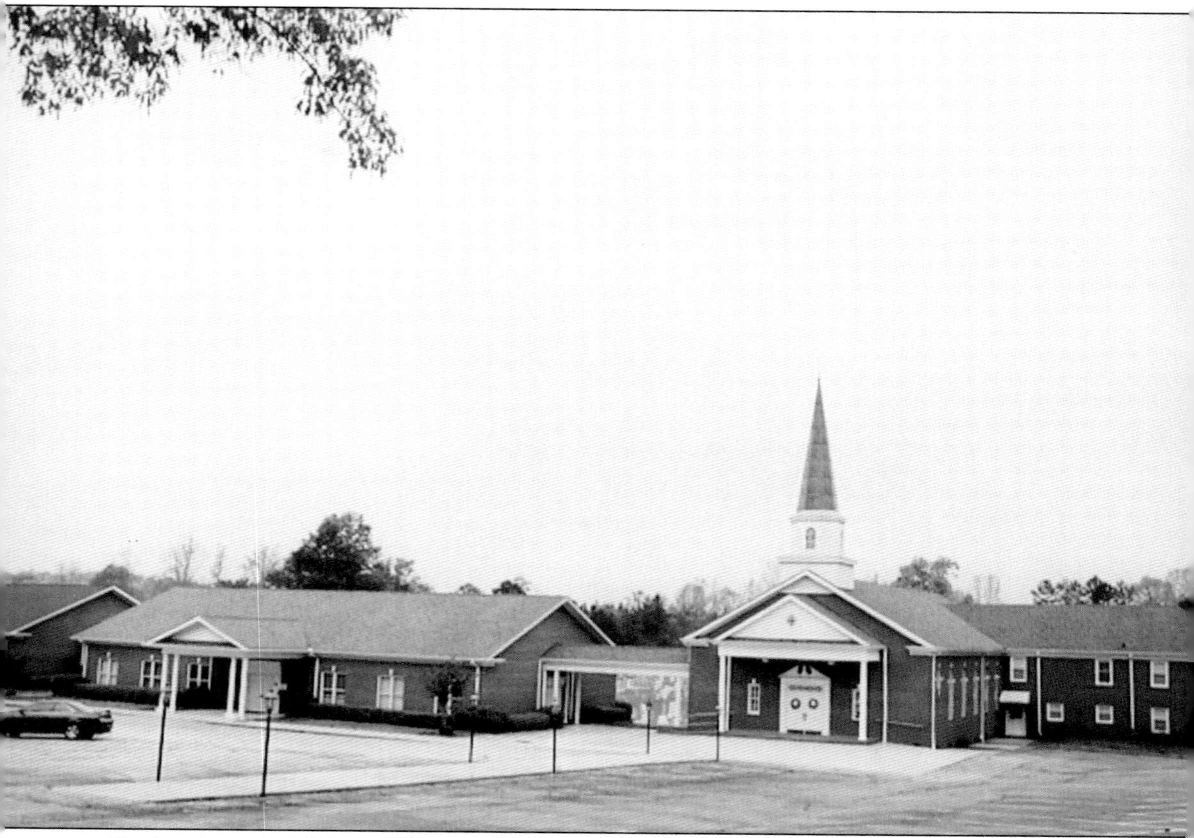

During the early 1900s, Sunday school classrooms were built, and portals were added to the church entrance. Additions and improvements have been made through the years, particularly the new auditorium-educational building completed in 1961. Fork Shoals Baptist Church has been the strong foundation for the community for over 200 years. (Courtesy of Angela Cox Photography.)

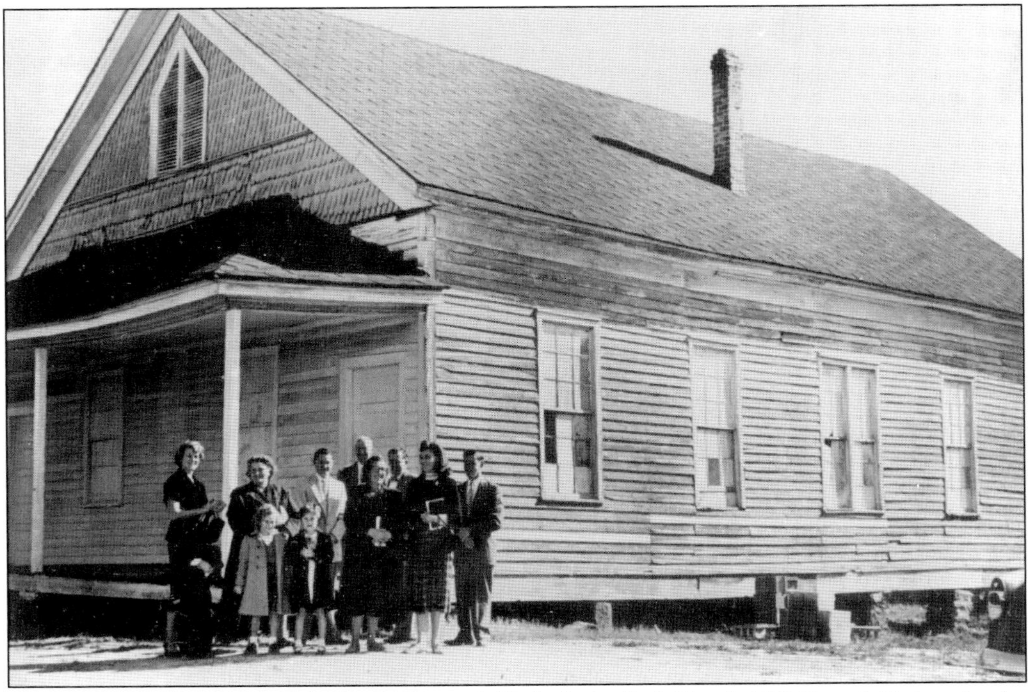

Will Nesbitt, owner of the Fork Shoals Cotton Mill, built the Fork Shoals Mill Chapel. He bought the old Pisgah United Methodist Church building after it erected its new chapel in 1904. He dismantled the old building and used the lumber to build a chapel near his mill. This mill chapel was used for various purposes through the years, including conducting nondenominational services, holding classes for the first private high school in 1907, and as a family dwelling. Services were again being held here, and in 1951, the Fork Shoals Mill Baptist Church was organized. Services were held here until a church was built on its present-day site nearby. The 1954 photograph below shows Pastor Roy Julian (right) receiving the deed to build the new Fork Shoals Mill Baptist Church. Later, the name was changed to Cedar Falls Baptist Church. (Courtesy of Nellie R. Thompson and Gloria Hughey.)

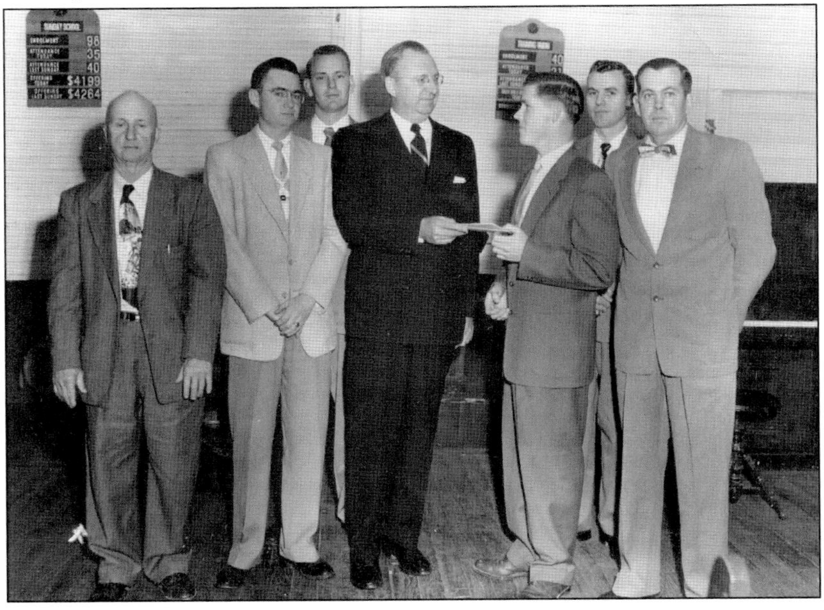

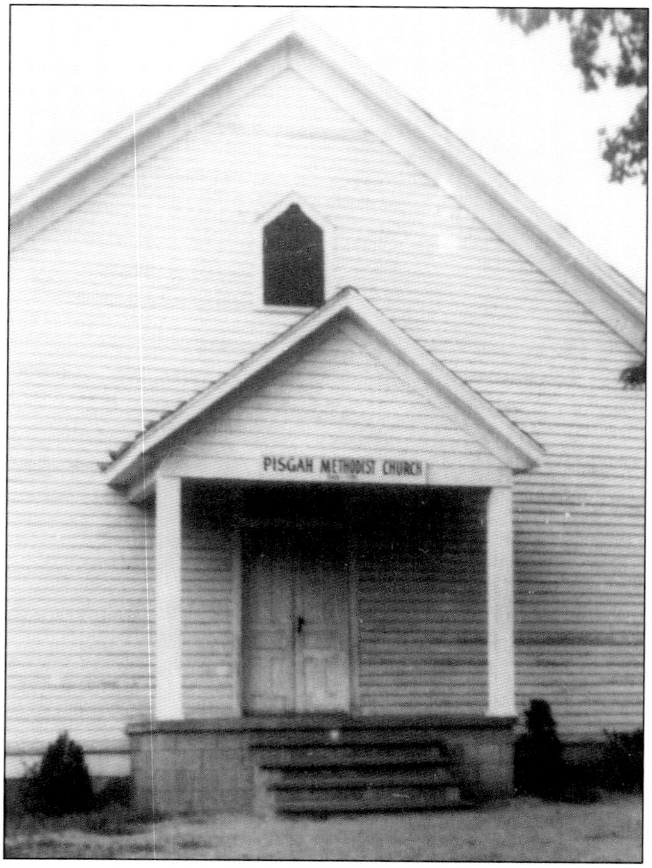

Thomas Terry established Pisgah United Methodist Church in 1791. It was originally called Terry's Chapel. It was made from hewn logs, chinked with mud, and had oiled paper windows. About 50 years later, a larger church was built nearby. In 1904, this larger wood-clad building was completed. It contained one large sanctuary, sectioned off with curtains for Sunday school classes. Pisgah United Methodist is one of the oldest churches in the upstate. Significant renovations to the 1904 structure have taken place. The exterior was bricked, a basement was dug for a fellowship hall, and additions were completed for use as educational wings to the church. Presently, it has a new fellowship hall with a kitchen and pastor's study constructed adjacent to the sanctuary. (Left, courtesy of Edwin Terry; below, courtesy of Angela Cox Photography.)

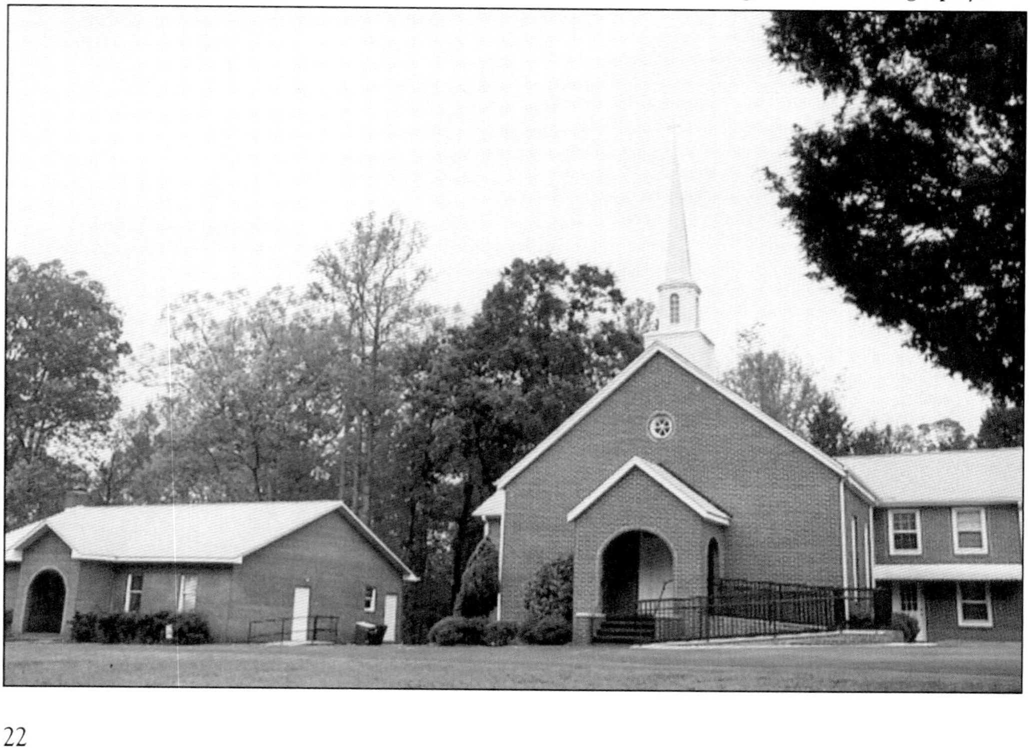

After the Civil War, in 1865, blacks began holding prayer meetings under a brush arbor on the Trowbridge Place, owned by Dunk Moore, later known as the Terry Place. In 1868, a church was organized under this brush arbor. It was first called Pleasant Grove and later renamed New Pleasant Grove when a second church building was erected in 1882 just southeast of Fork Shoals. (Courtesy of Virginia Agee Dean.)

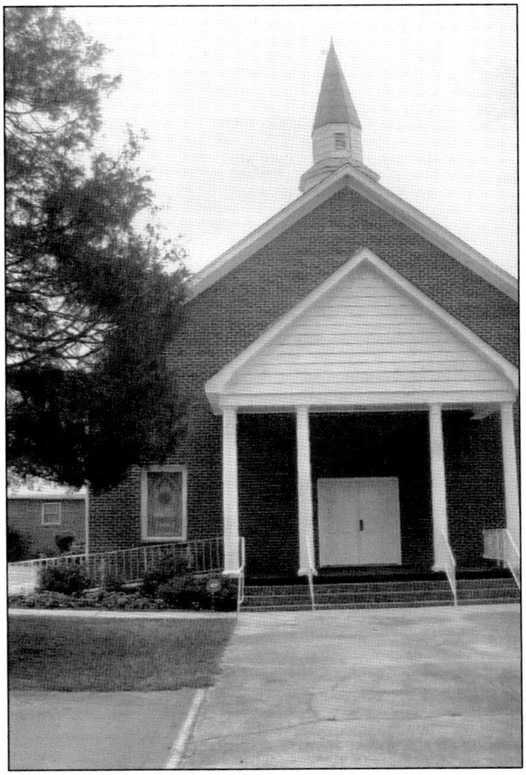

New Forksville Baptist Church started as a Sunday school in 1881, lead by Brother James Shumate. The following freed slave families incorporated the church in August 1884: Evans, Stewart, Perkins, McCullough, Gaines, Bolden, Butler, Hoods, Nash, Griggs, and Shumate. Three different sanctuary buildings have been worshiped in, and there is currently a Family Life Center. New Forksville is off Augusta Road near Princeton. (Courtesy of Virginia Agee Dean.)

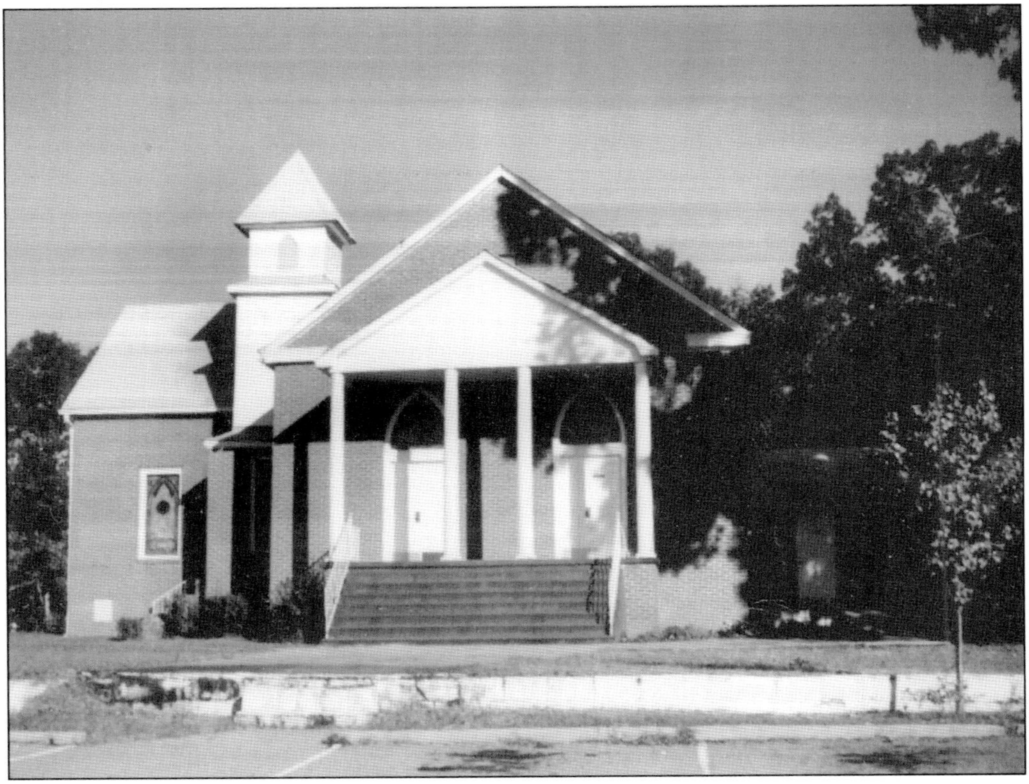

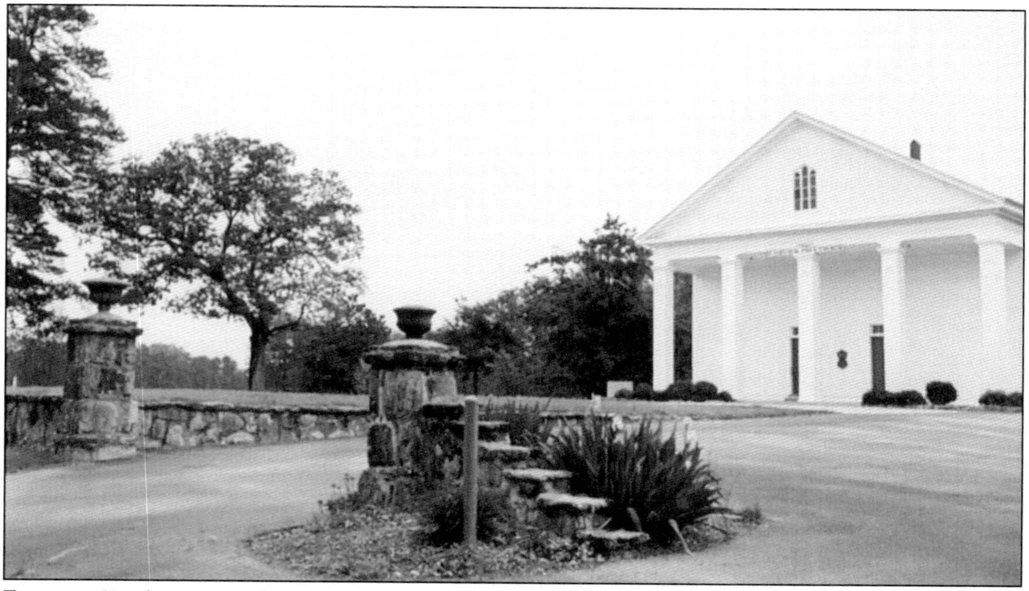

Fairview Presbyterian Church, founded in 1786, is the mother of Presbyterianism in Greenville County. Its original members hailed from Scotland via Ireland. The stone steps on the side of the driveway pillar are reminiscent of when ladies and others used them to climb into their carriages. Some traditions remain, and still today, deacons pass quaint pouches attached to rods to collect the offering. (Courtesy of Angela Cox Photography.)

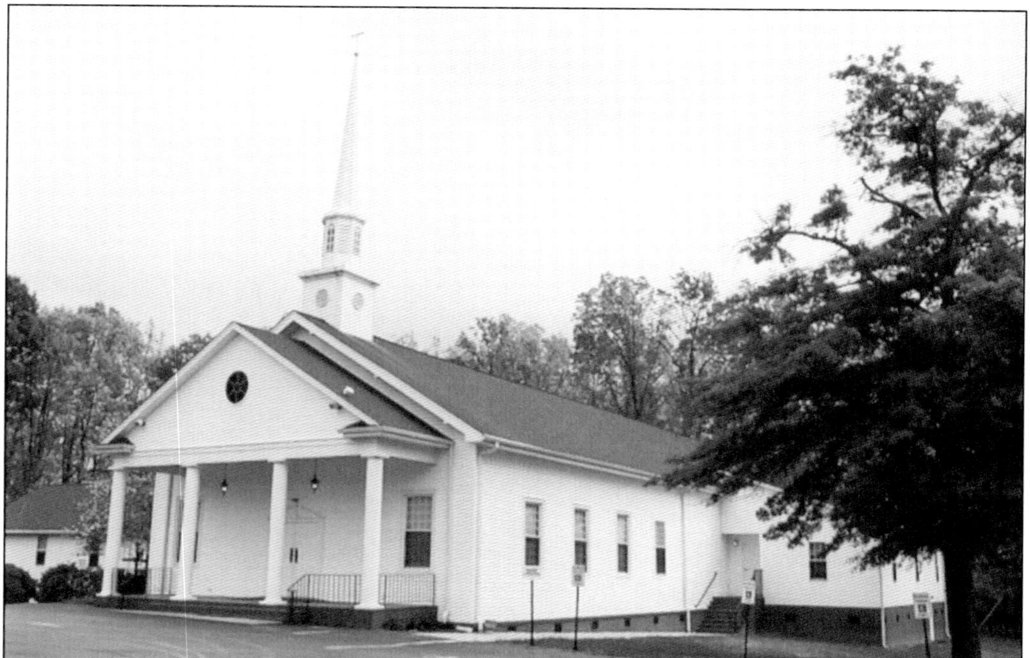

Hillside Baptist Church was established in 1903. People in the community built a multidenominational chapel in 1902, but only the Baptists came to use it. Oil lights were used for years, purchased by the women of the church with proceeds from selling hens. By about 1940, electric lighting and indoor plumbing modernized the church. Air-conditioning with gas heat was installed to replace the handheld fans and coal-burning stoves of the past. (Courtesy of Angela Cox Photography.)

Lebanon United Methodist Church, founded in 1785, was the first Methodist church in Greenville County. Originally a log cabin, or "pole chapel," it served as a church and school. It was called "the Grove" because it was on Charles Sullivan's land thus named. In 1832, its location was moved to higher ground nearby and renamed Lebanon. The late William D. Sullivan had papers showing that James Cowan was contractor and received payments for the meetinghouse. Building committee members were Benjamin Camp, Abijah Pinson, and William Meares. Years later, elders could remember services in "the old plank church." The present church building was constructed in 1852. Modernizations and remodeling occurred during the first half of the 1900s. The church built a parsonage in 1959. (Both, courtesy of Jim Scott.)

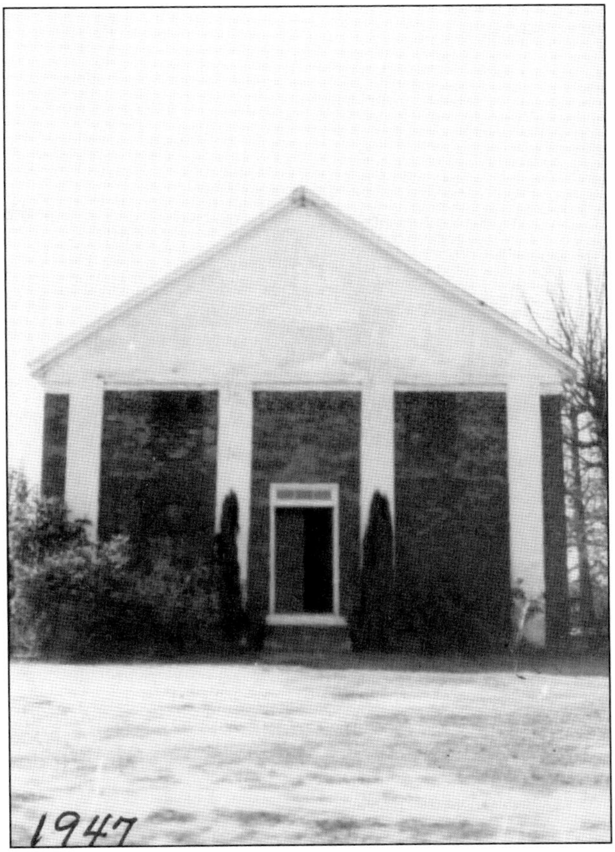

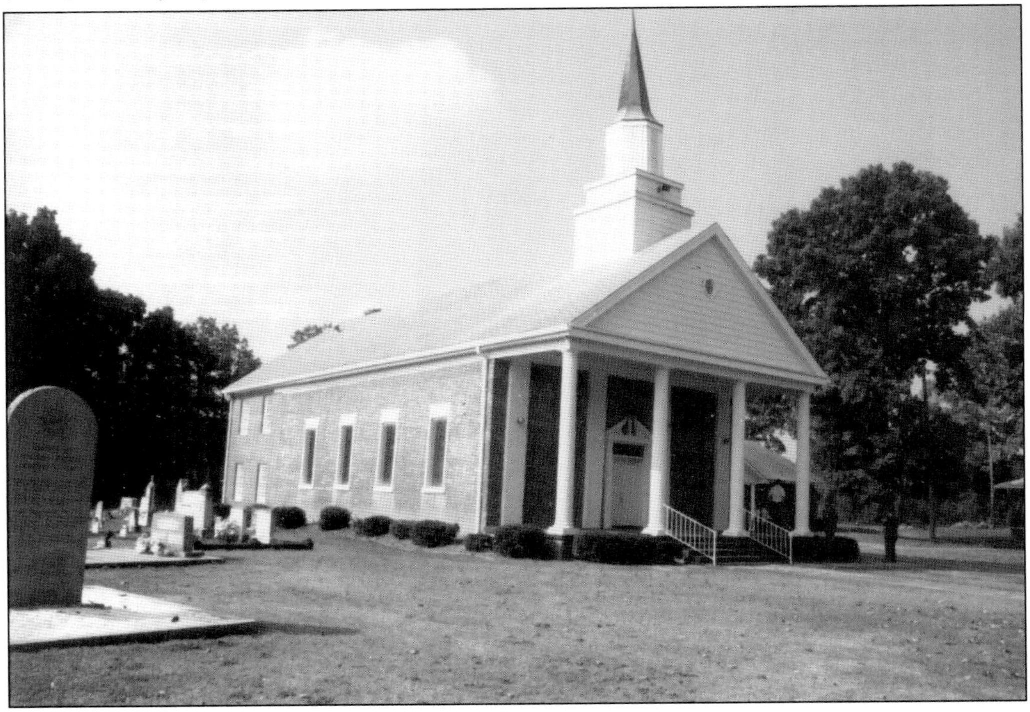

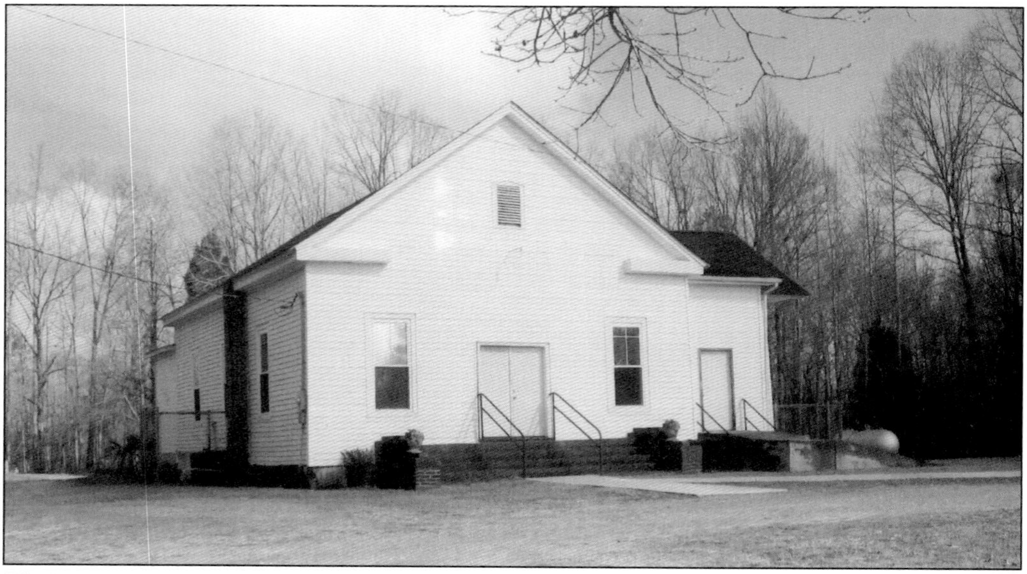

The New Prospect Baptist Church was established in 1867 and is believed to be one of the oldest churches in South Carolina established by and for freed slaves. William Henry Pool donated the land and materials; many of the first members were current or former workers at the Pool farm. The original building is still used today with some additions and improvements added over the years. (Courtesy of Jim Scott.)

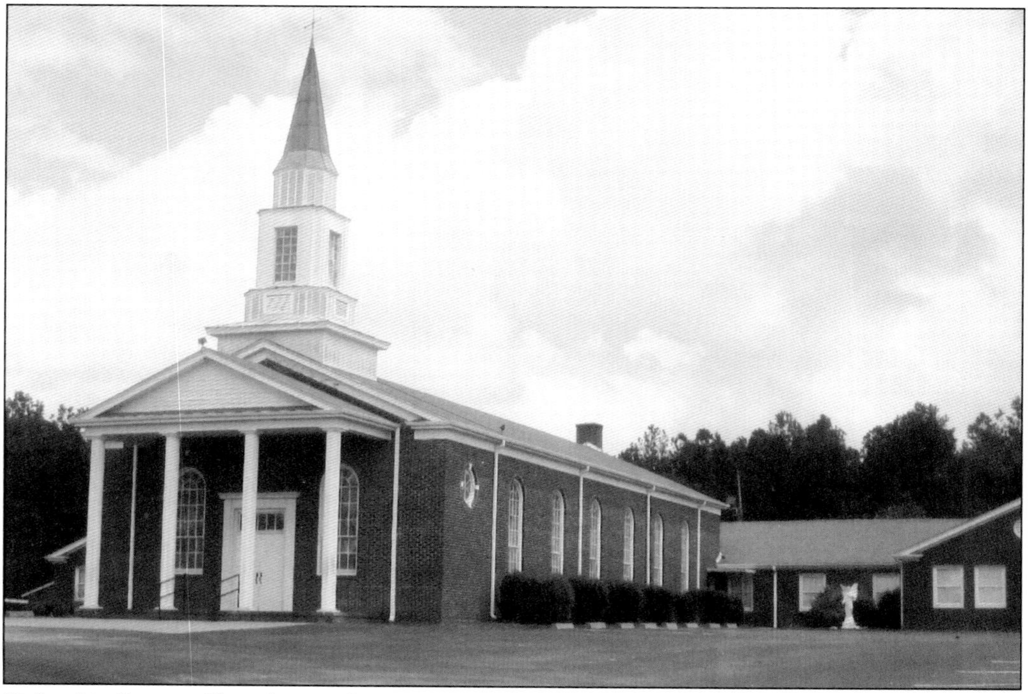

Columbia Baptist Church near Princeton was organized in 1815 as an offshoot of Fork Shoals Baptist Church. It is unknown where the name originated; however, it is speculated that it could have been due to the fact that when people traveled Highway 25 from North Carolina to Columbia, South Carolina, this church marked the halfway point. There was an annual call of pastor up until 1952, when the church eliminated that policy. (Courtesy of Virginia Agee Dean.)

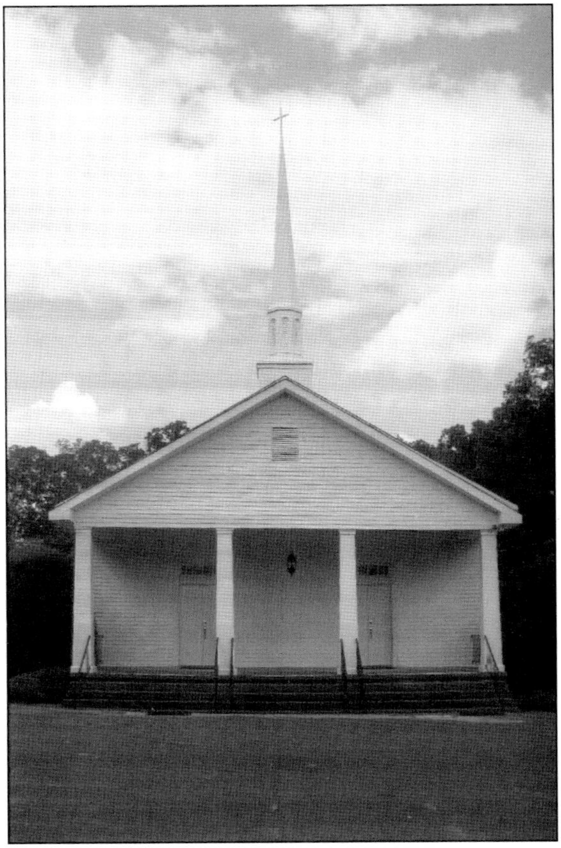

Oak Hill United Methodist Church was established as an arm of Washington Baptist Church at Horse Creek Academy in 1838. In 1842, an application was granted to move to the present location at Oak Hill. Shortly thereafter, a circuit rider of the Methodist church was permitted to preach, and through his ministry, they decided to embrace the Methodist denomination. It was named Bethesda Methodist Church until 1934. (Courtesy of Virginia Dean and Lena Horton.)

Washington Baptist Church off Highway 8 was organized in 1821 in the kitchen of John Sullivan as an arm of the Fork Shoals Baptist Church and became independent in 1824. The origin of the name is unknown. When the black membership became freedmen, they were dismissed so they could have their own churches. The present-day Shady Grove Baptist Church is the church they formed. (Courtesy of Linda Hufstetler and Virginia Agee Dean.)

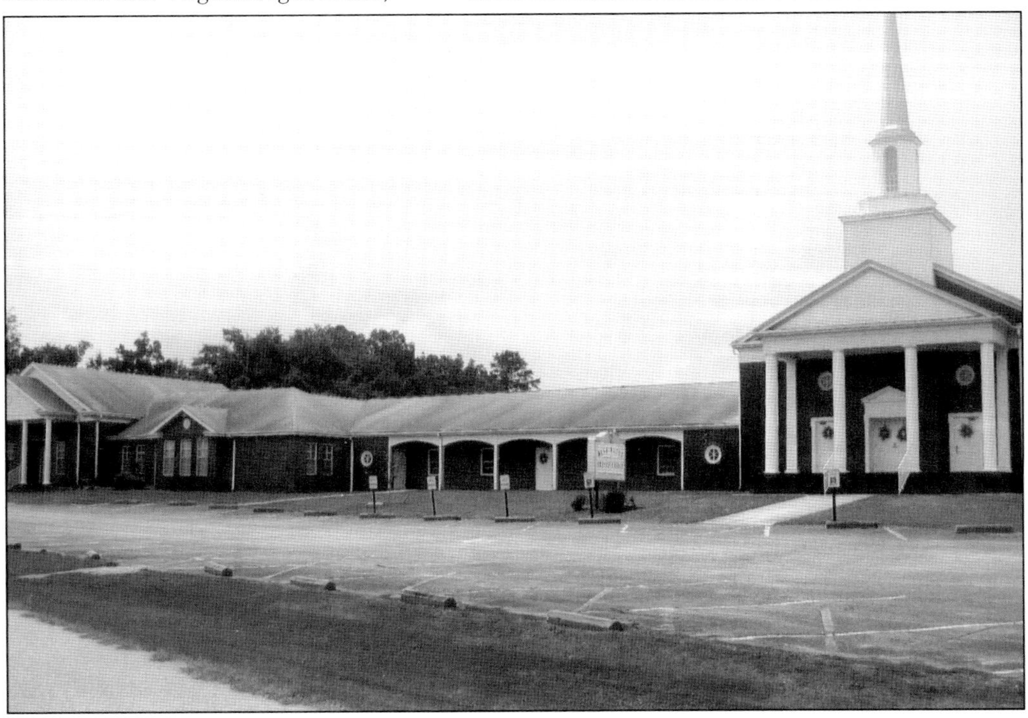

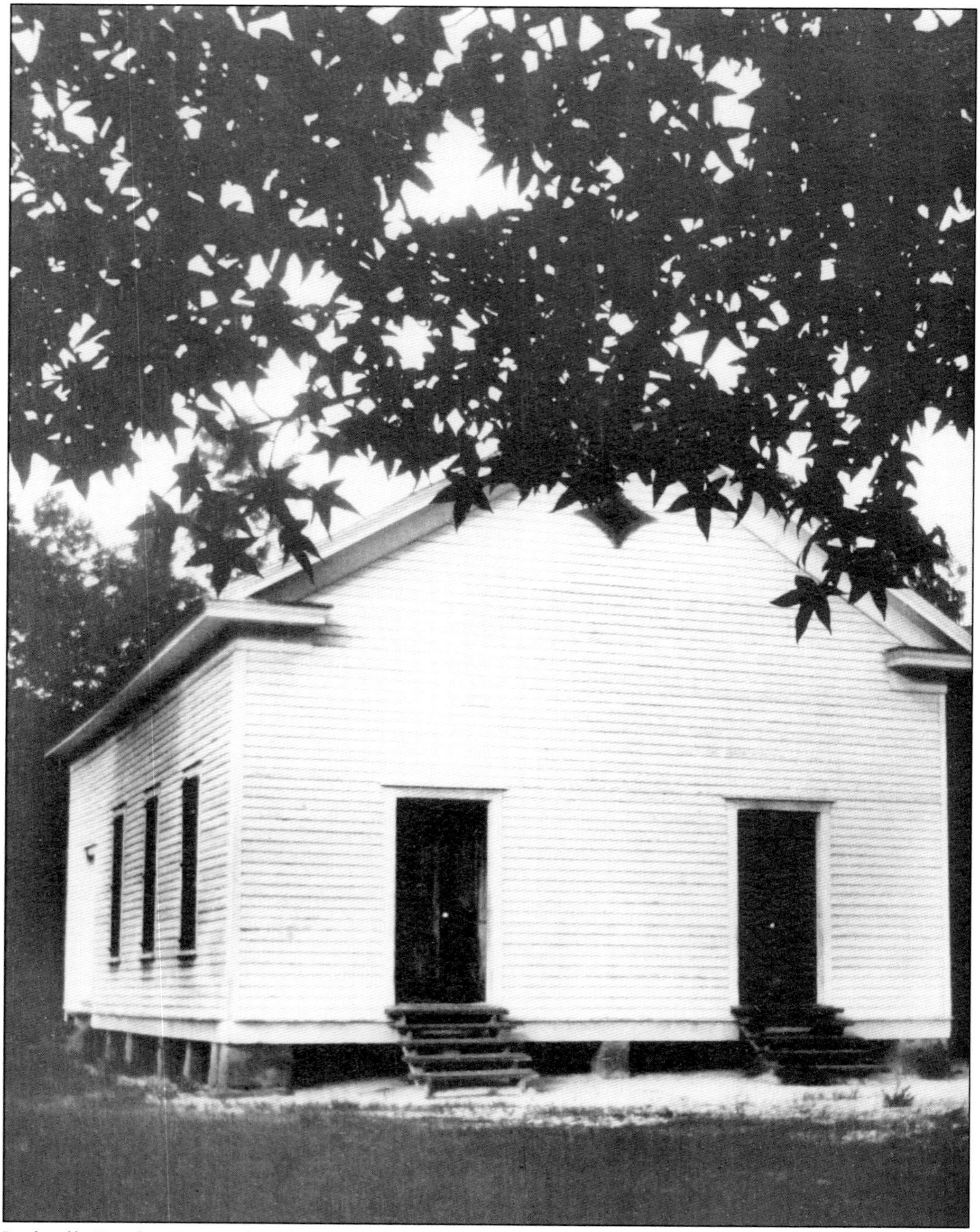

Lickville Presbyterian Church was founded in 1882. Rev. C.L. Stewart and a commission from the Enoree Presbytery formally established the church with 20 charter members. The first services were organized on Ellen C. Woodside's property in a small schoolhouse. She was determined to have a place of worship established in her community. Rev. C.L. Stewart would stop by to preach on his way between Fairview and Piedmont on horseback, and a congregation of neighbors began to organize. (Courtesy of Lickville Presbyterian Church.)

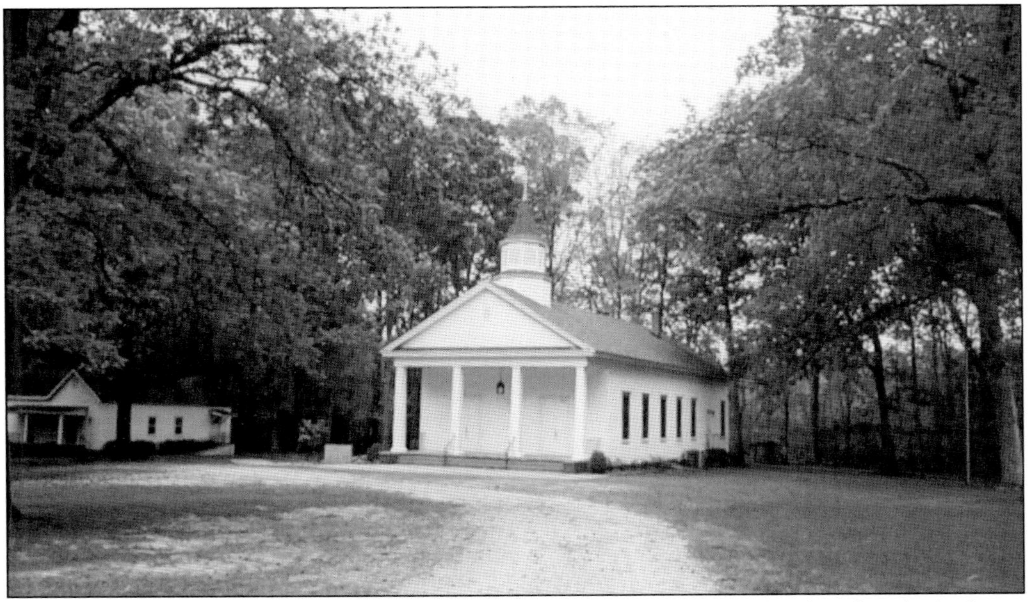

W.A. McKelvey provided a site on Augusta Road below Ware Place for the church to be built not far from his estate. This frame building, constructed in 1882, was remodeled 61 years later and again in 1968. A session house was also built in the 1880s and was used as a school. The church cemetery includes plots of many prominent families from the Lickville community. (Courtesy of Lickville Presbyterian Church.)

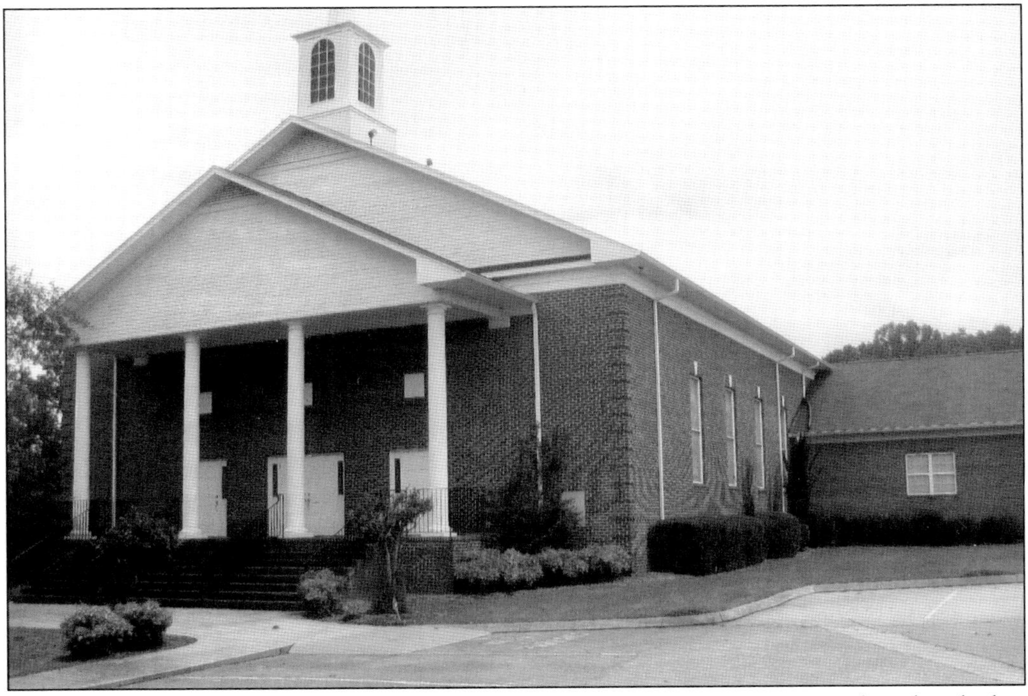

Shady Grove Baptist Church near Ware Place, now a large congregation, is another church that formed after the Civil War. The church was established in 1867 by 14 black men and women who set out from Washington Baptist Church. One and a half miles away, they started worship services under a brush arbor led by the Rev. Peter Shorter. (Courtesy of Virginia Agee Dean.)

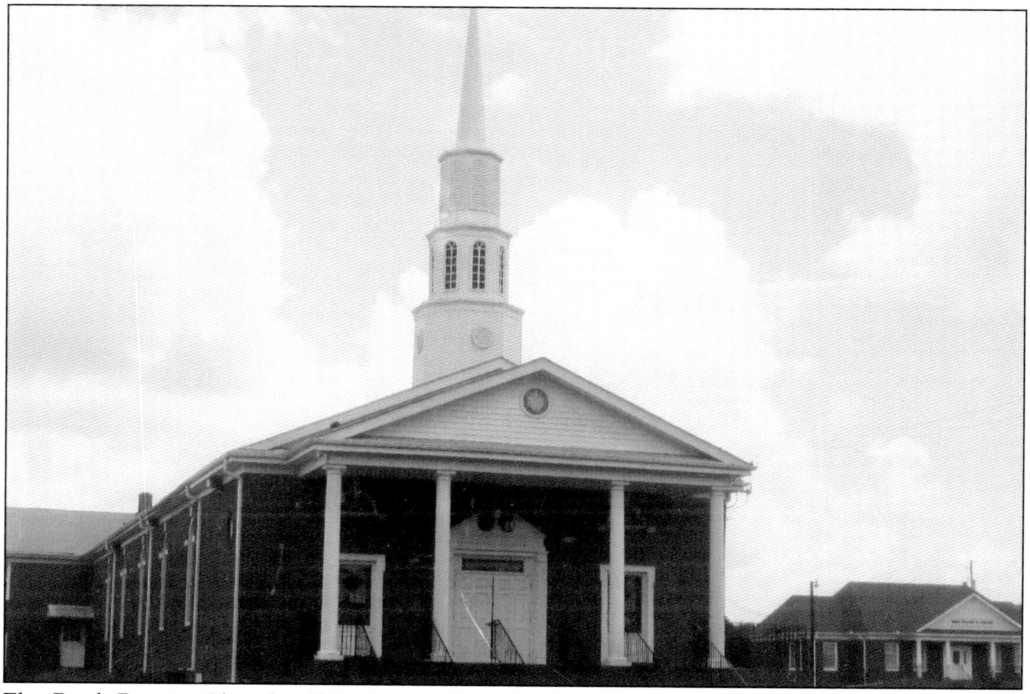

Flat Rock Baptist Church off Highway 86 was organized in 1862 and met on a large flat rock in a brush arbor. This is where the church got its name. In 1890, the church bought three acres of land from W.B. Charles for $60 and constructed a building. Flat Rock is a thriving, growing church. (Courtesy of Virginia Agee Dean.)

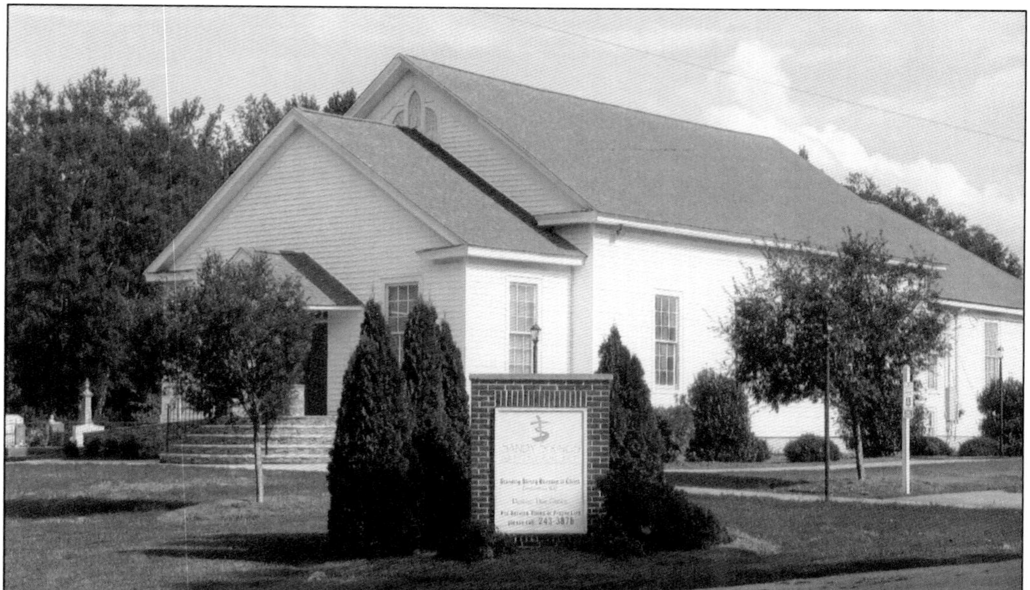

Sandy Springs Baptist Church near Woodmont High School was established in 1832 as an offspring of Standing Springs Church near Simpsonville. The congregation initially met in homes. Following the Civil War, church members constructed the present sanctuary, but the exact date is unknown since records from 1862 to 1897 were lost. Sandy Springs is a small but welcoming congregation. (Courtesy of Anne Peden.)

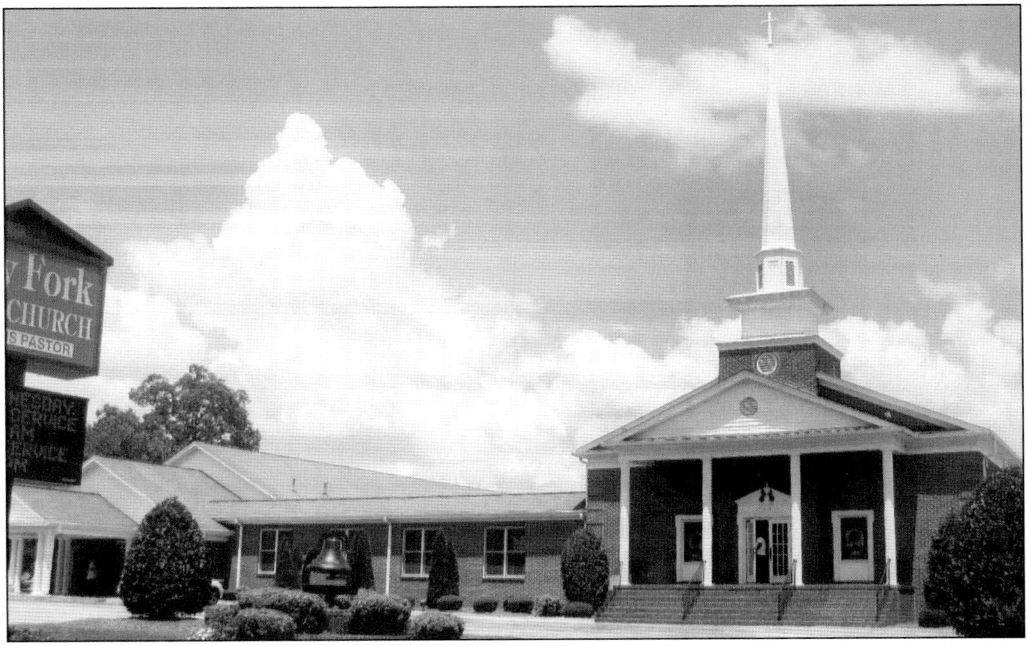

Reedy Fork Baptist Church on Fork Shoals Road began worship in 1850 in a log cabin lead by Rev. Hugh Sullivan. The official organization was in 1868 and was incorporated in 1896. A brush arbor structure was built on land donated by Newton Sullivan in 1872. The colored fairgrounds were behind Reedy Fork Church in the early 1900s. It is a vital, growing church now. (Courtesy of Virginia Agee Dean.)

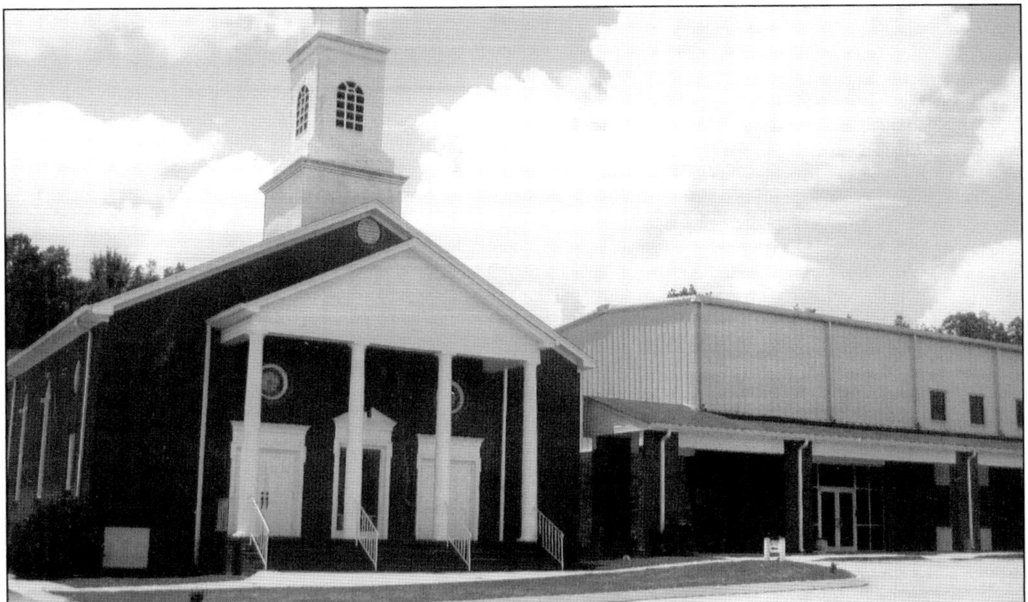

In 1895, Dolfus Sweeney founded Rocky Creek Baptist Church, located near the Simpsonville Campus of Greenville Tech, by meeting in the home of the Thackstons. As the congregation grew, it moved to a vacant house in a section called Hammett Town. In the spring of 1896, fifteen members were baptized. The first church was built with donated lumber. Now, Rocky Creek is a large congregation of believers. (Courtesy of Virginia Agee Dean.)

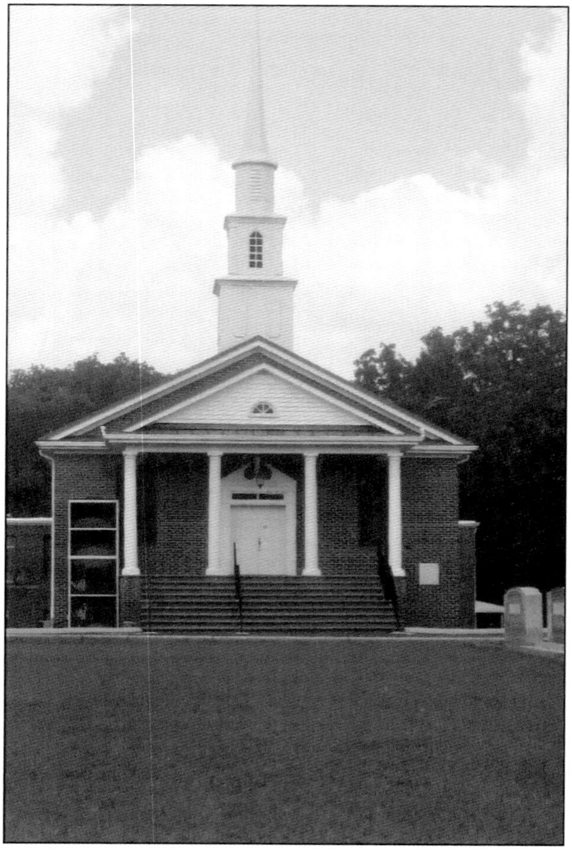

Princeton Baptist is just west of Augusta Road in the Princeton area on Highway 76 toward Honea Path. (Courtesy of Virginia Agee Dean.)

Daventon Baptist Church was organized in the early 1900s when members holding letters of dismissal from other churches met at the old Davenport graveyard. Twelve folks in all formed the new congregation, and the first pastor was Rev. J.H. Machen. The following year, R.L. Davenport donated the land to build Daventon Baptist Church. (Courtesy of Virginia Agee Dean.)

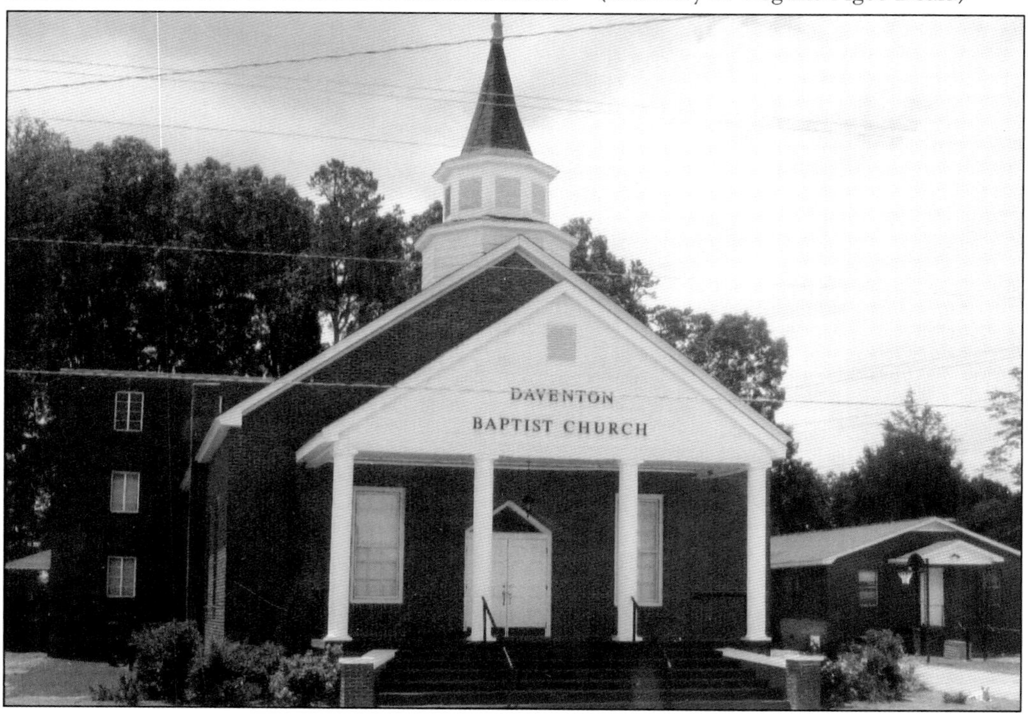

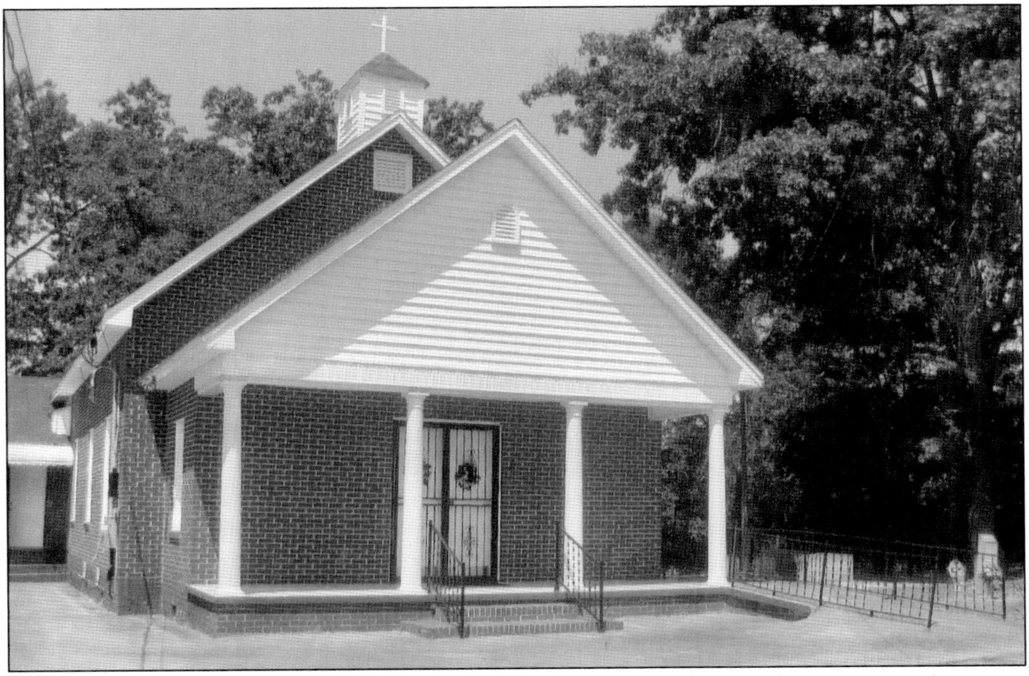

Holly Springs Baptist Church had its beginning in 1880, when the members met at a place near a spring and a holly bush, which was called a brush arbor. After land and lumber were donated, the congregation worked hard to construct a wooden-frame building the same year. As with many nearby churches, the school building was used as the first meeting hall. The school building, shown below, stills stands near the current church. The Holly Springs School building is the oldest African American school building in Greenville County. (Both, courtesy of Virginia Agee Dean.)

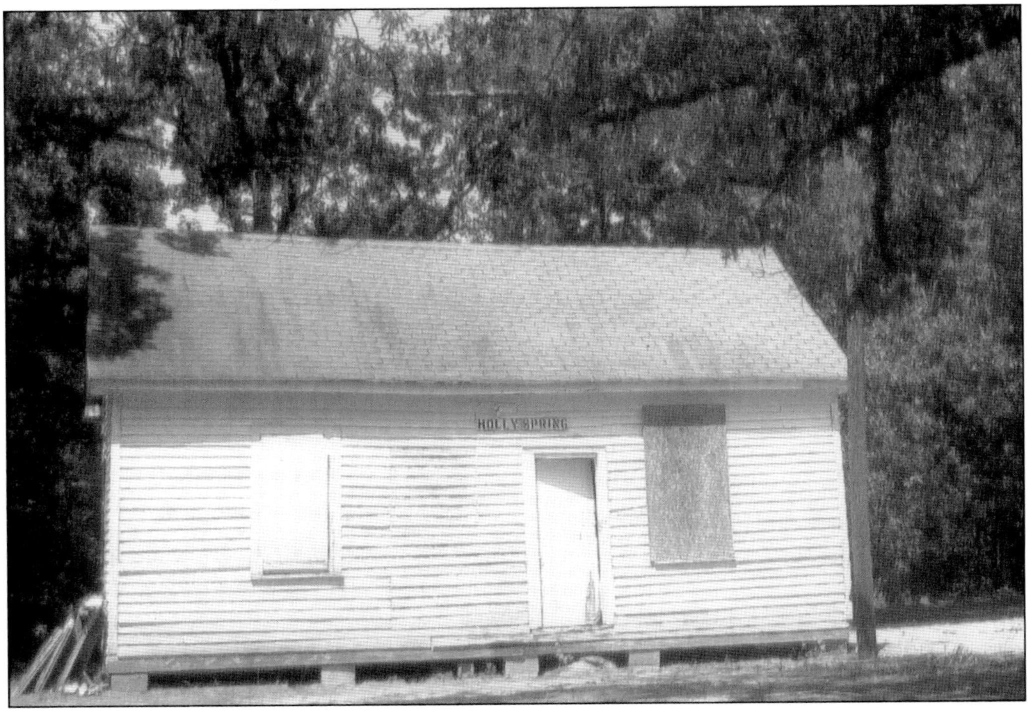

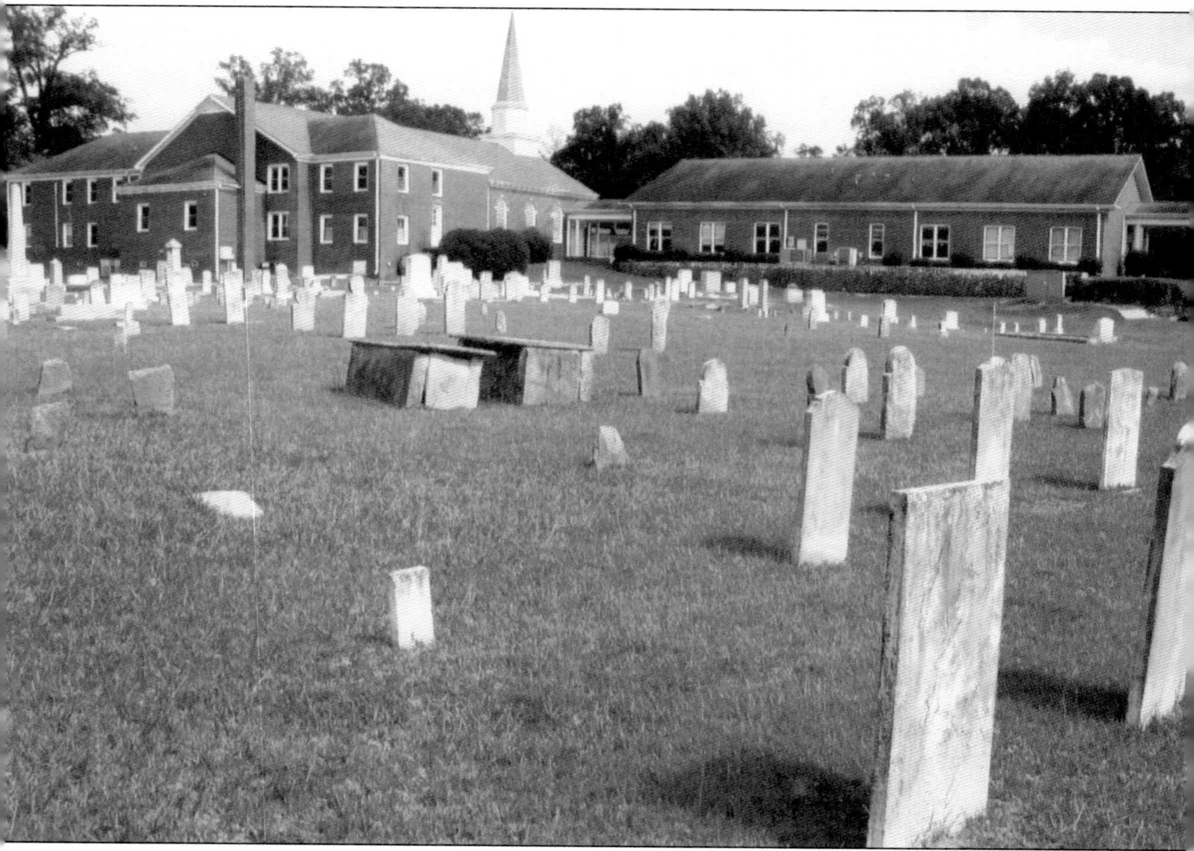

This view of Fork Shoals Baptist Church is from the graveyard. One of the oldest cemeteries in the area, Fork Shoals contains the graves of Hudson Berry and his wife, Sarah Anthony. Hudson was one of the earliest settlers in the area and owned large tracts of land as well as several businesses at the Cedar Falls. Many graves date to the early 1800s, as evidenced by the markers and aboveground graves in the foreground. (Courtesy of Anne Peden.)

Three

SOLID FOUNDATIONS

SCHOOLS

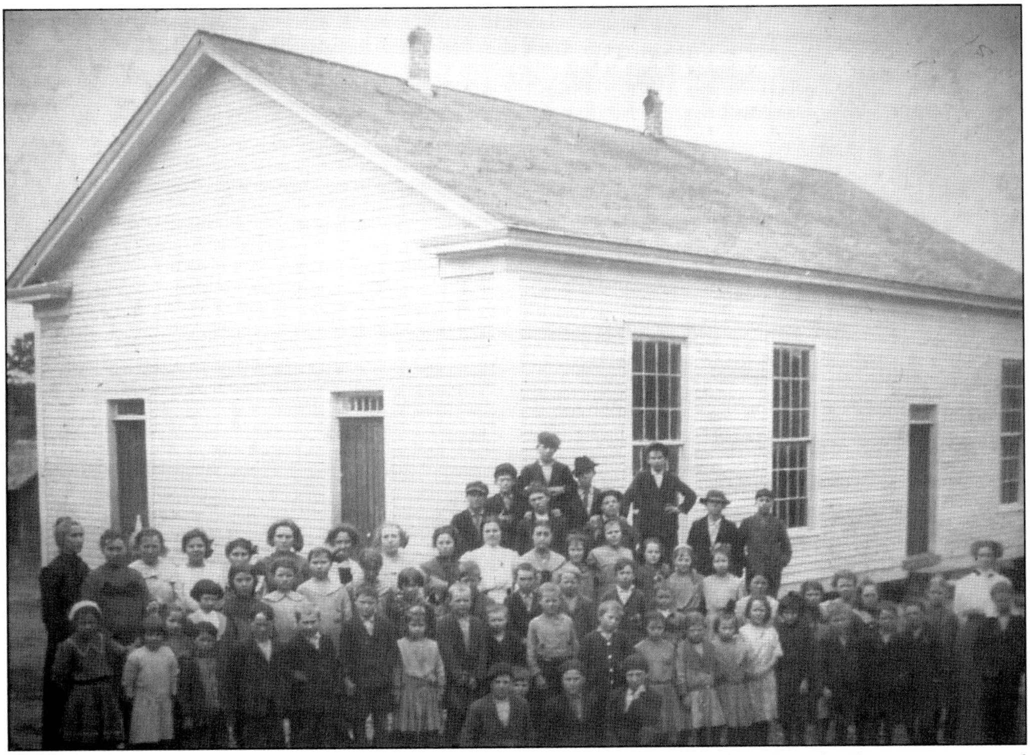

Fork Shoals School had its beginnings with Fork Shoals Baptist Church, which was organized between 1777 and 1780. The school was reportedly founded about 1877 but likely functioned with the church from its inception. There were two school buildings during the 1800s on the church grounds. The second one had one large room downstairs and one upstairs used by the Masons and Woodmen of the World. (Courtesy of Cheryl MacKnight.)

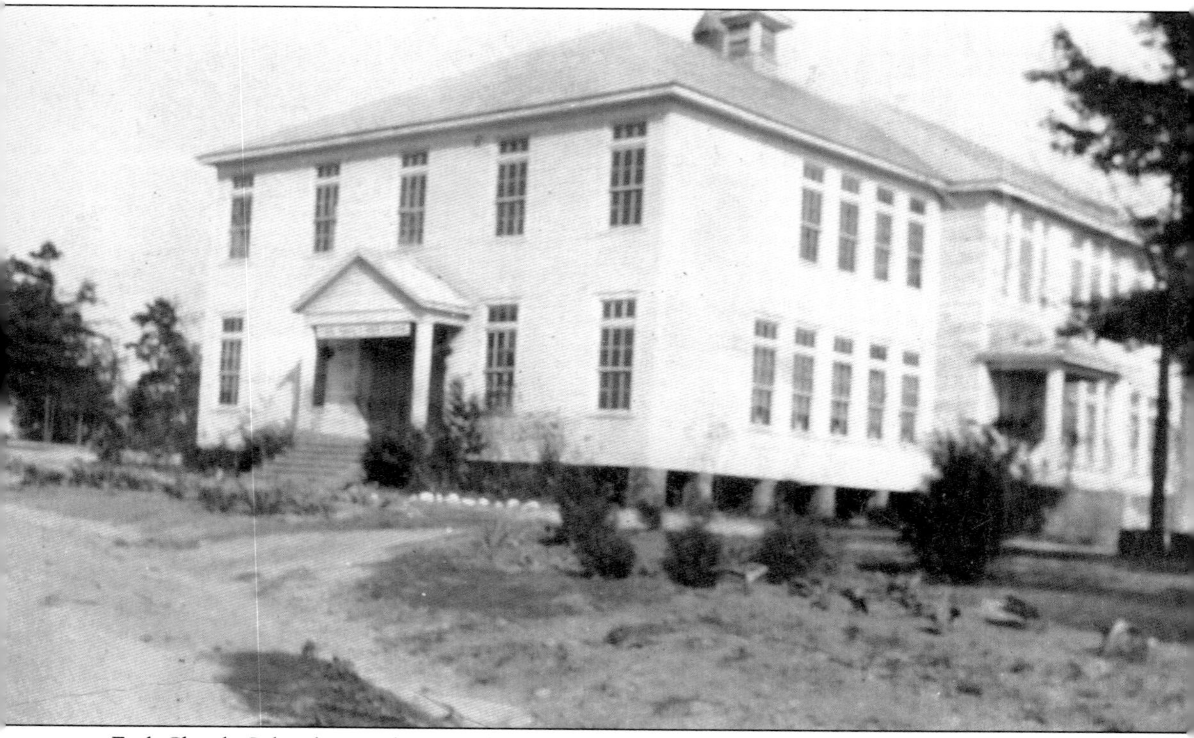

Fork Shoals School moved across the road from the church into a new building in 1916. It had two large classrooms downstairs and an auditorium upstairs. It was located on land donated by Dr. W.A. Ross and Frank Austin. In need of more space from 1920 until 1923, high school grades were taught in the old private high school building constructed in 1907–1908 on the east side of the Reedy River, while the lower grades remained put. For these three years, the school was operating on both sides of the Reedy River. Fork Shoals School became a state high school in 1922. One year later, this large addition was completed, with the upper grades returning. A new three-building school complex was constructed in 1939. Its lunchroom was located on the ground floor of this 1916 building. (Courtesy of Cheryl MacKnight.)

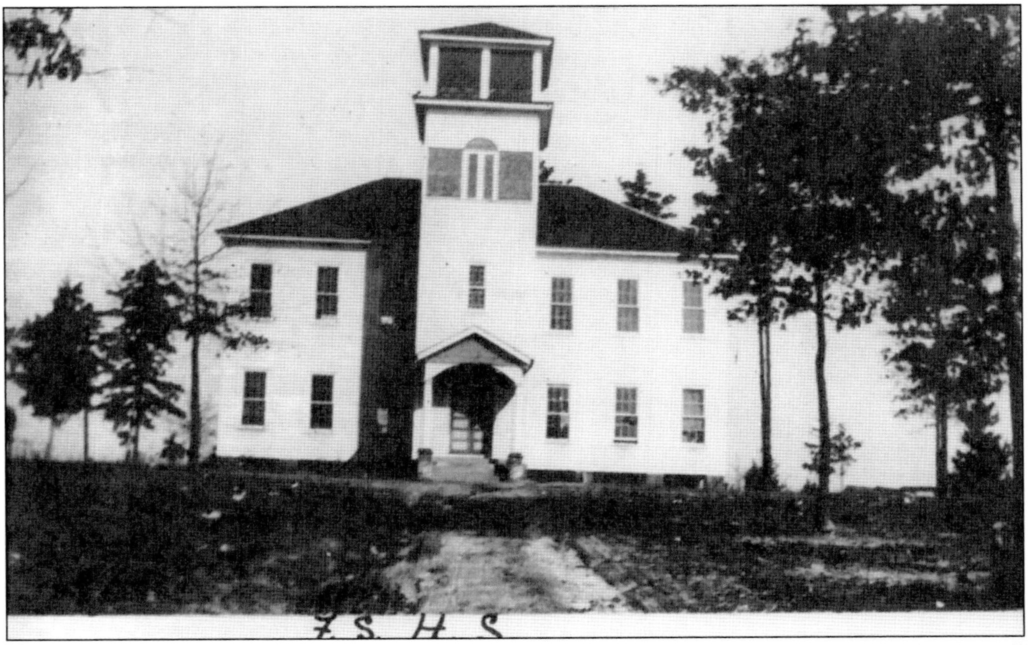

Located on a hill just east of the Reedy River on property obtained from the Fork Shoals Mill and completed in 1908, Fork Shoals Private High School operated as a private school until 1917. During World War I, teachers were hard to find, young men were called to service, and free public schools in nearby communities all influenced the decision to close the private school. (Courtesy of Cheryl MacKnight.)

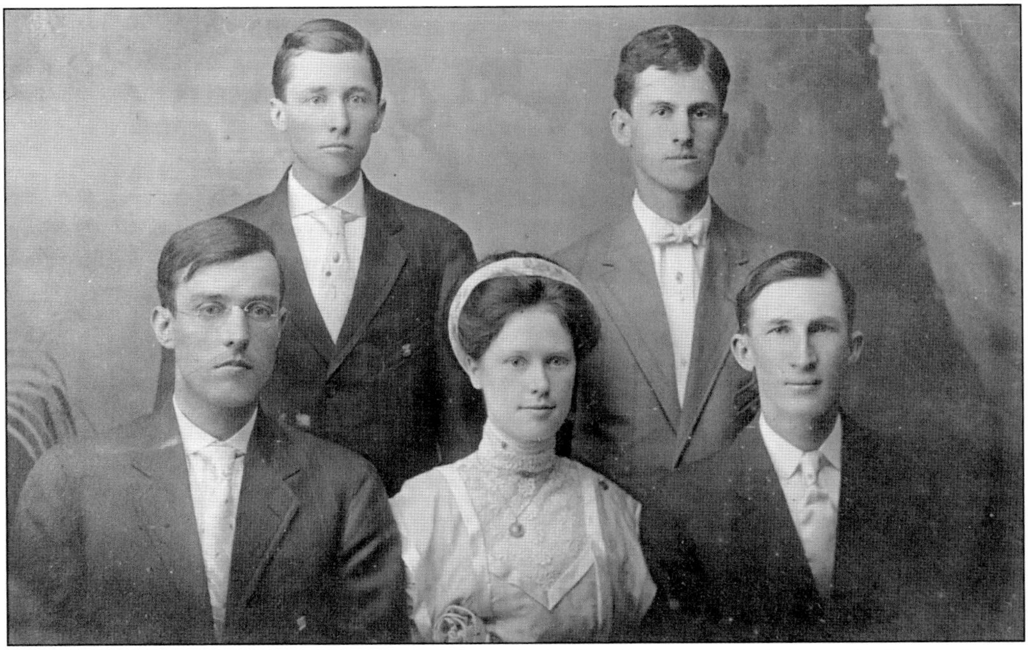

The class of 1910 was the first to graduate from Fork Shoals Private High School. The graduates were, from left to right, (first row) Clinton P. Rice, Katie Scott, and Clifford Stewart; (second row) S.H. "Nick" McKittrick and William A. Hopkins. Rice returned to the school and served as superintendent of the high school from 1922 until 1930. (Courtesy of Cheryl MacKnight.)

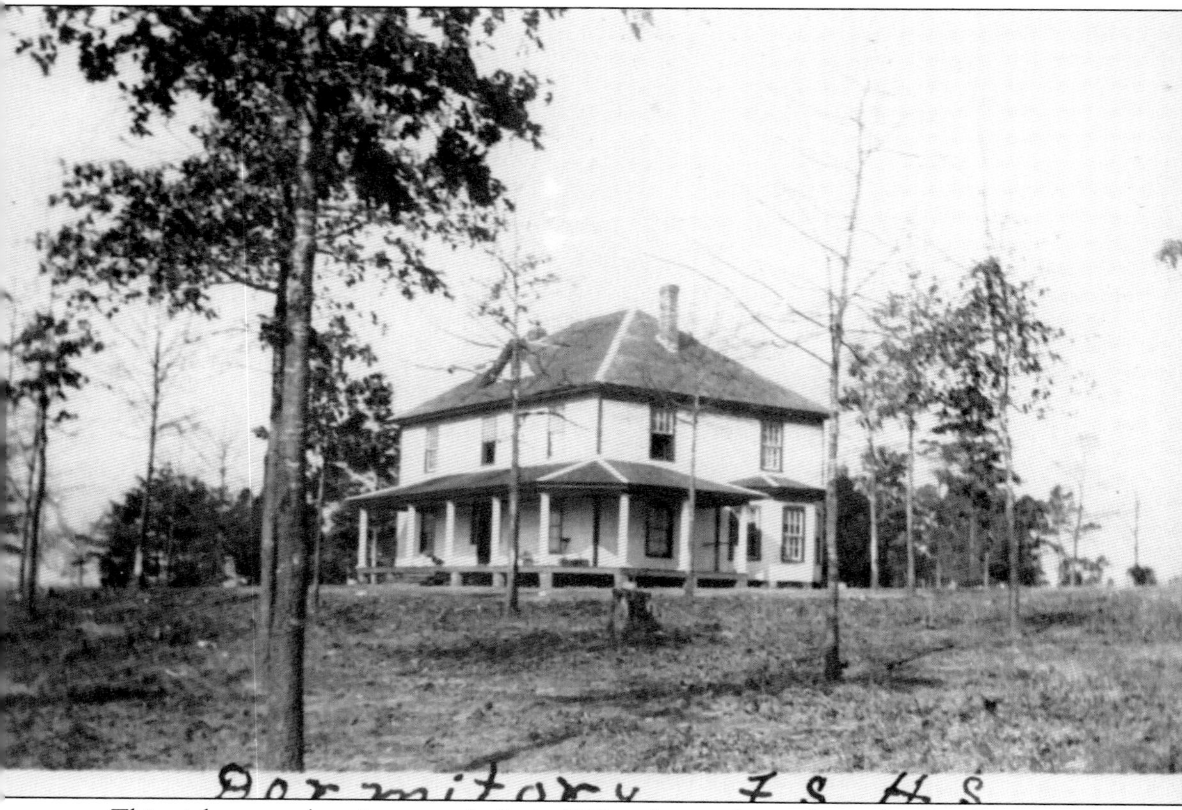

This eight-room dormitory was built in 1911 to house the professor, his family, and female students. It was located just to the west of Fork Shoals Private High School, both on the east side of the Reedy River. The student body moved into a new building on the west side of the river in 1916, the same location as today. The private school closed the next year; although, its doors were reopened temporarily for three years, from 1920 until 1923 for the upper grades, for need of classroom space. Once the addition was completed, the high school students returned from across the river. Sometime after this, the property of the private school was sold. The dormitory became a private residence for many years and was still standing in the 1970s but burned down sometime later. The old private high school building was eventually dismantled and hauled away to be used at the fairgrounds behind Reedy Fork Baptist Church on Fork Shoals Road. (Courtesy of Cheryl MacKnight.)

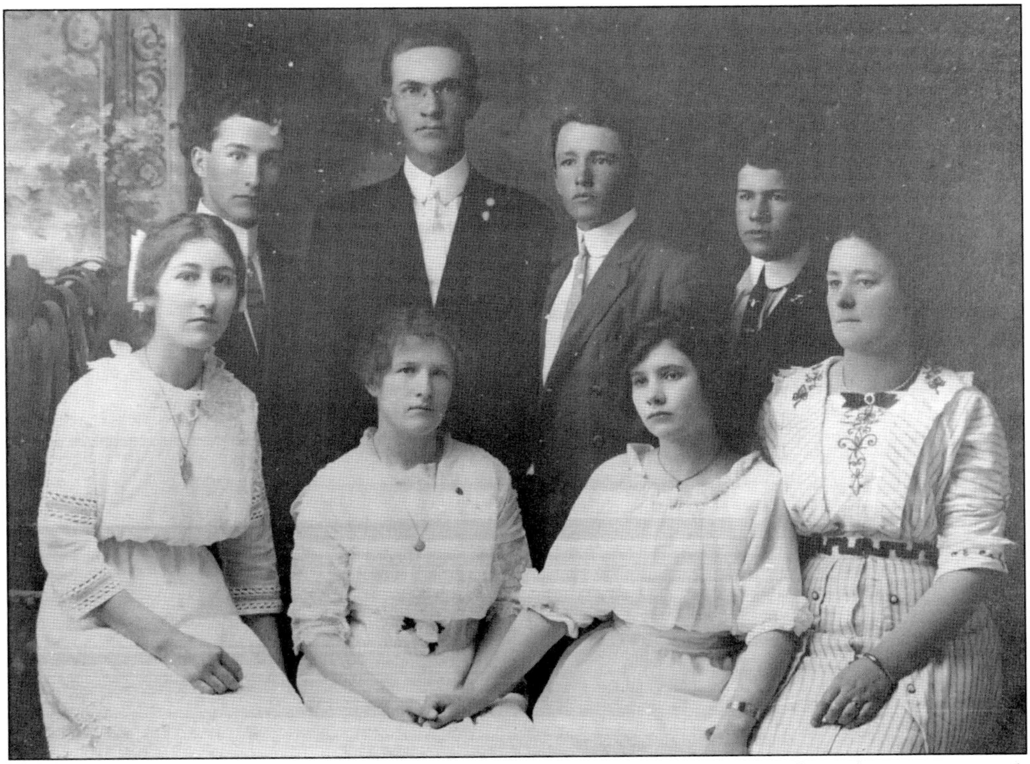

The Fork Shoals graduating class of 1914 consisted of 10 graduates; only eight are pictured. Prof. H.B. Jordan was the superintendent and teacher. The students' names were (from left to right) Blanche McCauley, Bonnie Davenport, Sarah Nash, Attrice Davenport, Mytie Jenkins (not shown), Thomas Goldsmith, Hoke Stewart, Lake McKittrick, Talmadge Crumpton, and Wofford Medlock (not shown). McKittrick was awarded a scholarship to the Citadel. (Courtesy of Cheryl MacKnight.)

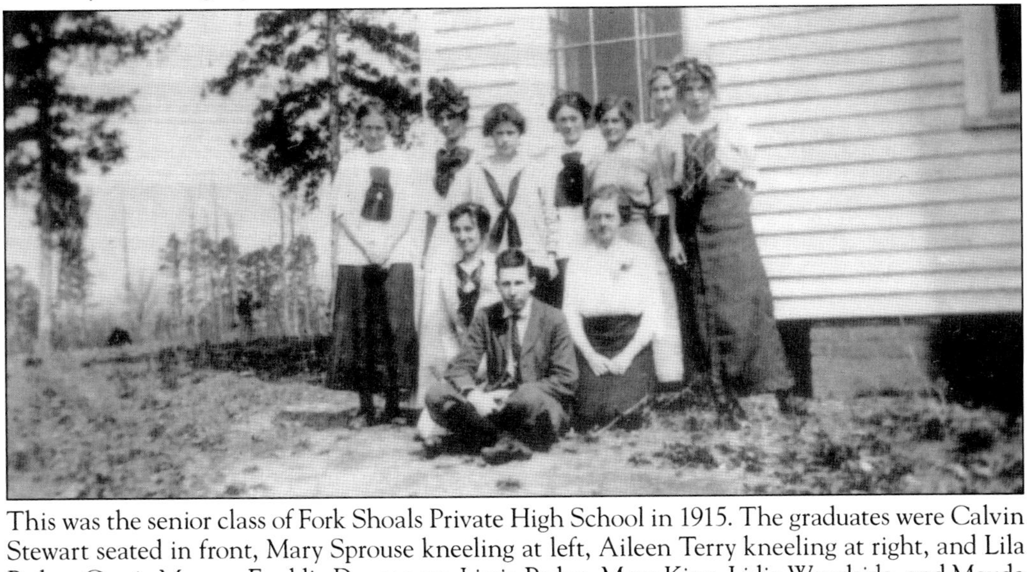

This was the senior class of Fork Shoals Private High School in 1915. The graduates were Calvin Stewart seated in front, Mary Sprouse kneeling at left, Aileen Terry kneeling at right, and Lila Peden, Carrie Meares, Freddie Davenport, Lizzie Peden, Mary King, Lidie Woodside, and Maude Woodside standing in no particular order. (Courtesy of Cheryl MacKnight.)

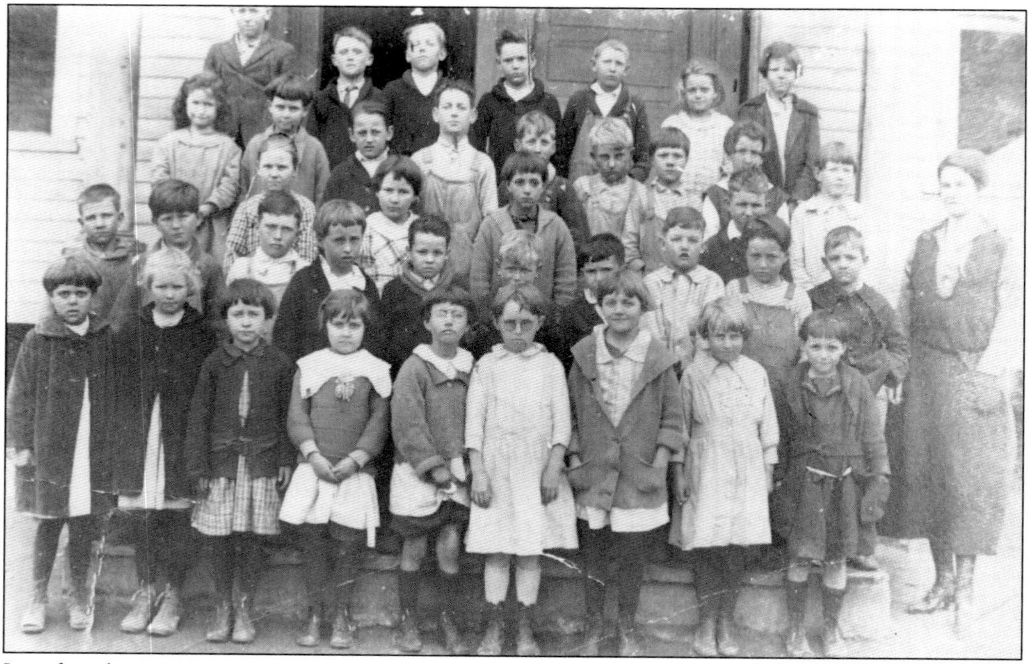

In a few short years, Fork Shoals School had grown from two teachers in 1916 to four teachers in 1919–1920. Miss Huff (above right) taught this class of young students during the 1922–1923 school year. Another teacher was added the following year, and a large gym was built in 1925. (Courtesy of Cheryl MacKnight.)

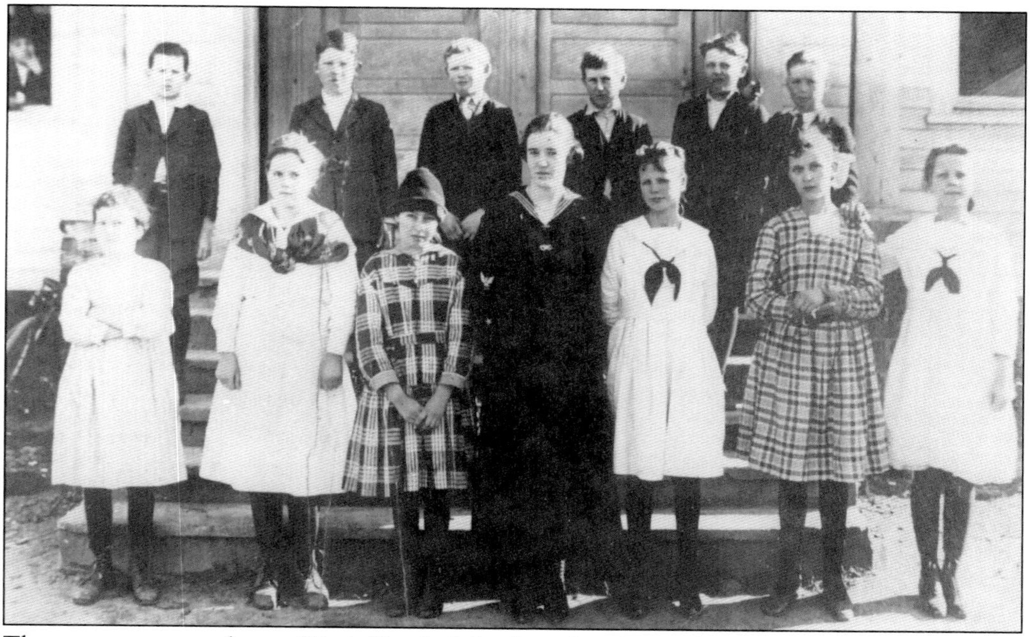

These are young students of Fork Shoals School in the 1920s. The teacher standing in front is Mrs. Rufus Coker, the former Myrtle Crisp. Between 1920 and 1923, the school was in need of more space, so the older high school students attended classes in the former private high school building across the Reedy River. The elementary school continues to grow to this day. (Courtesy of Cheryl MacKnight.)

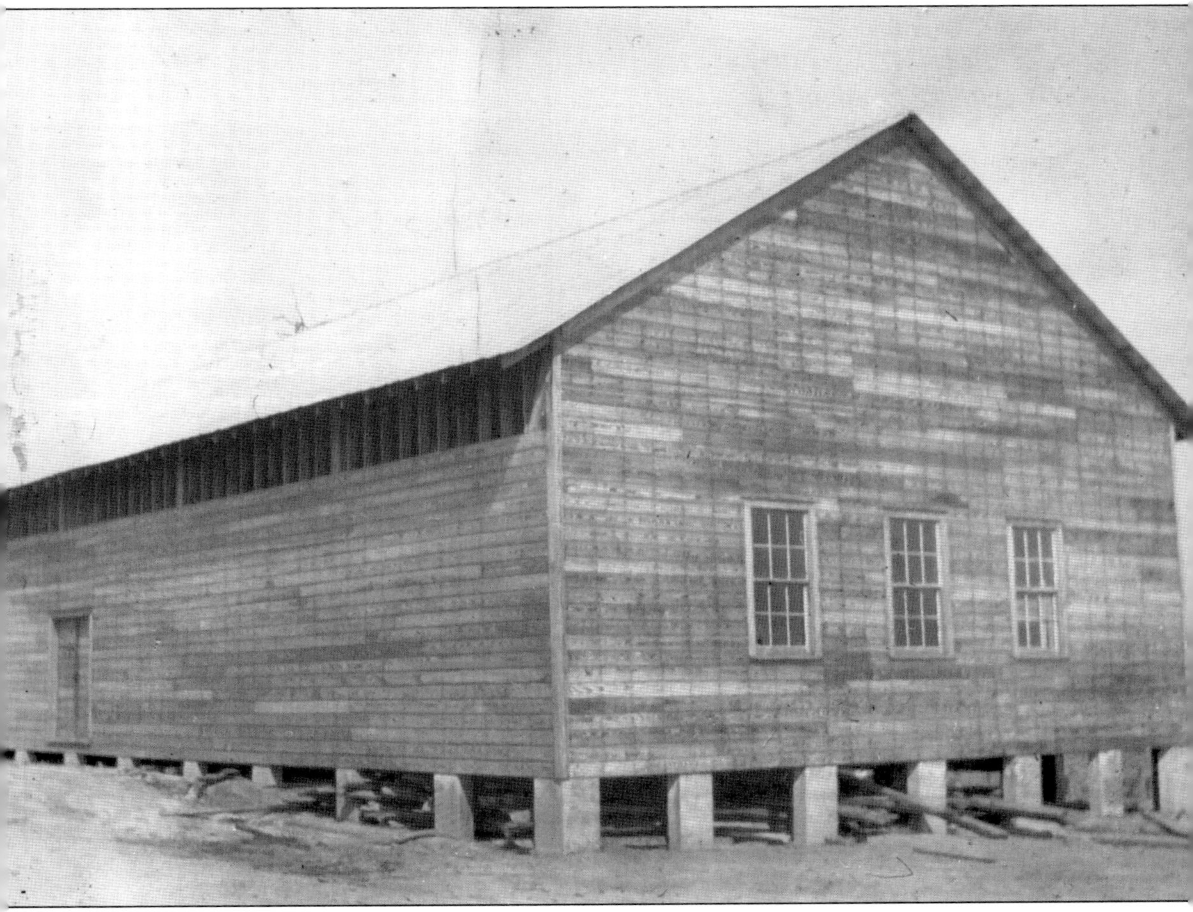

While Clinton P. Rice was superintendent of Fork Shoals High School, a gymnasium in back of the school was built in 1925, shown above. Spectators alternately hovered around an old stove to keep warm and shivered on the sidelines as they cheered on their team. There was a kind of balcony with no railing. Folks would climb up there and sit on the floor, swinging their legs over the edge. (Courtesy of the Peden family.)

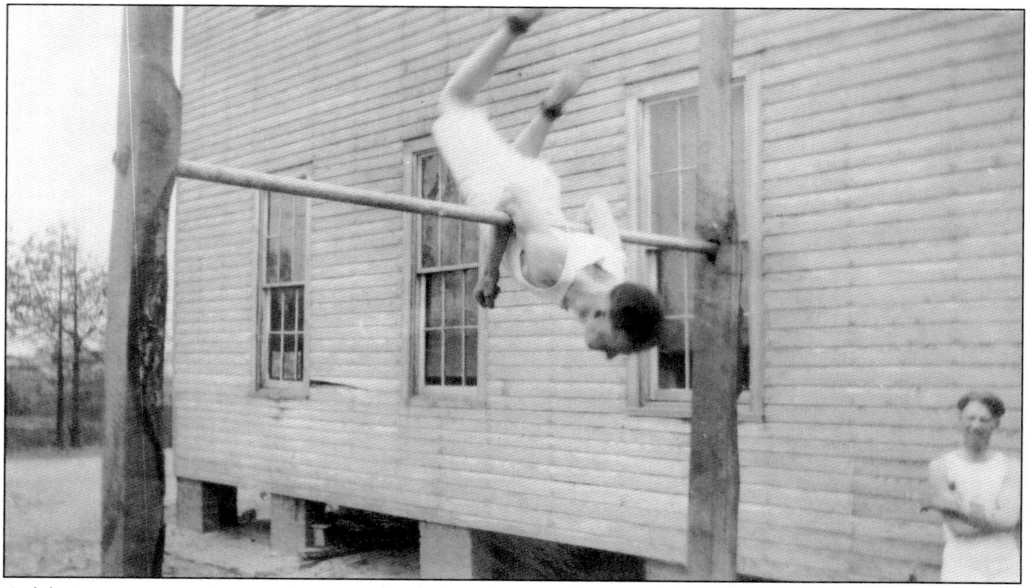

Athletics were a serious endeavor at Fork Shoals, and there was strong community support as well as involvement. Joe Peden is on the horizontal bar. The school's academic program excelled, and the Fork Shoals students competed successfully in a variety of district-wide scholastic contests. These same students competed in athletic events, beautification programs, musical programs, variety shows, and field days. (Courtesy of the Peden family.)

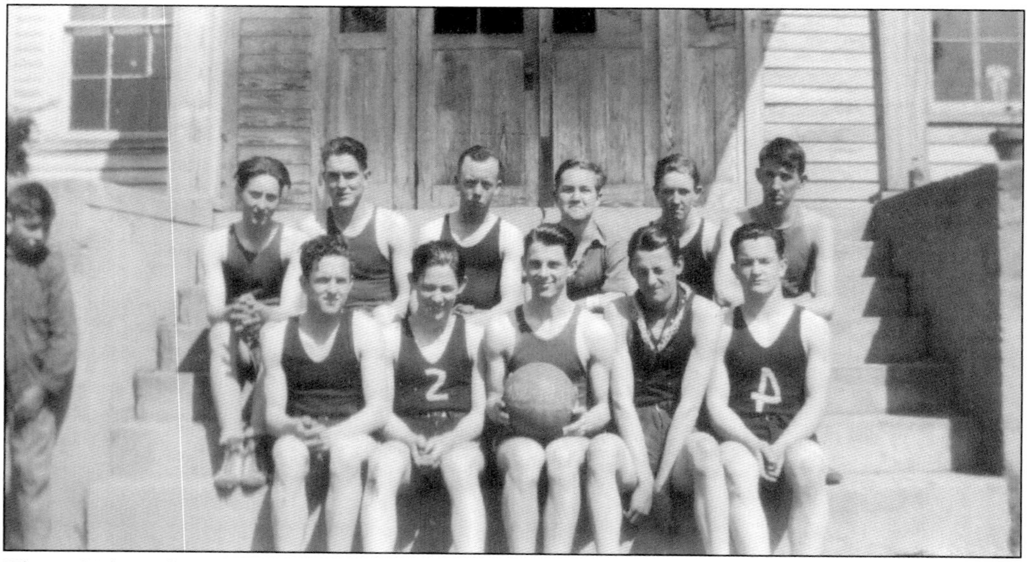

The 1933 boys' basketball team consisted of (in no particular order) captain Edwin Sims, Ernest Kellett, Leroy Trammell, Claude Willis, Frank Sims, Howard Thompson, Harrison Pruitt, and Charles Griffith, coached by W.C. Poore. Home games were played that year against St. Albans and East Gantt. Away games against Welcome, Mauldin, and St. Albans were played at Parker. Simpsonville, Berea, Traveler's Rest, and Fountain Inn were played on their courts. (Courtesy of Cheryl MacKnight.)

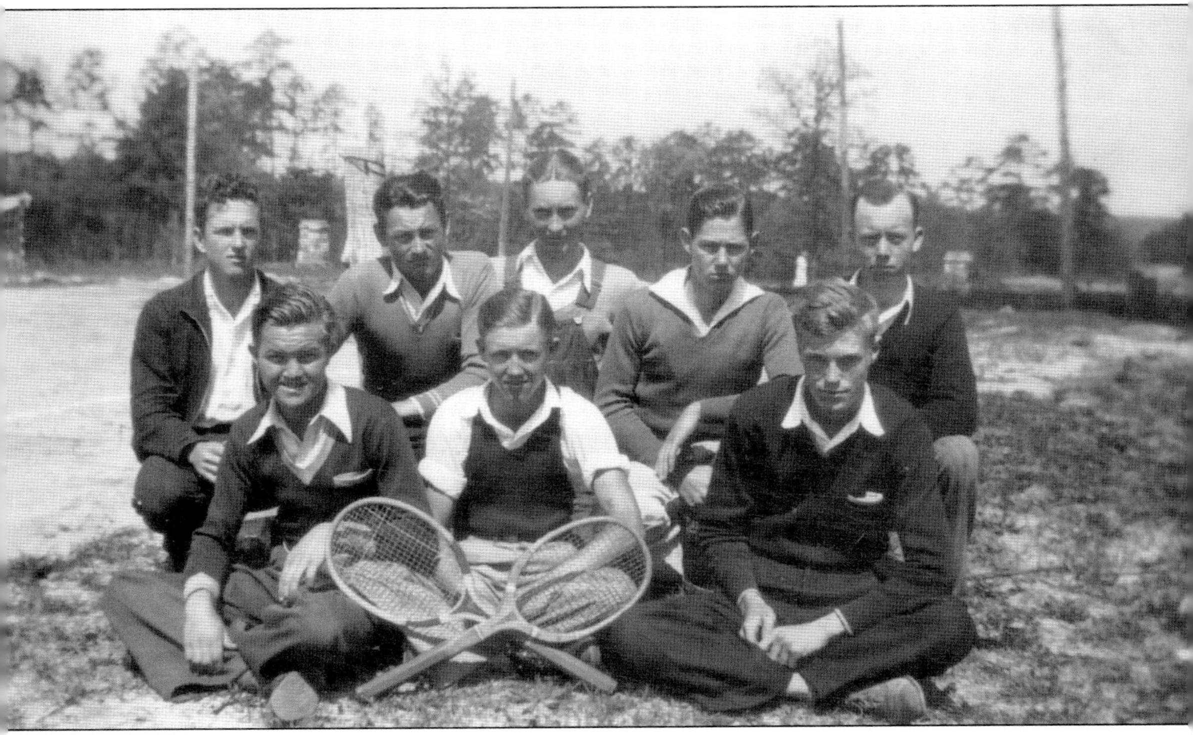

Fork Shoals School had a large number of activities available to students, including sports, music, and academic pursuits. These are the members of the 1933 tennis club of Fork Shoals High School. The members included Claude Willis (first row, center) and (in no particular order) Wilton Ridgeway, Leroy Trammell, Julian Hopkins, Charles Griffith, Frank Sims, Earnest Kellett, and an unidentified player. (Courtesy of Edwin Terry.)

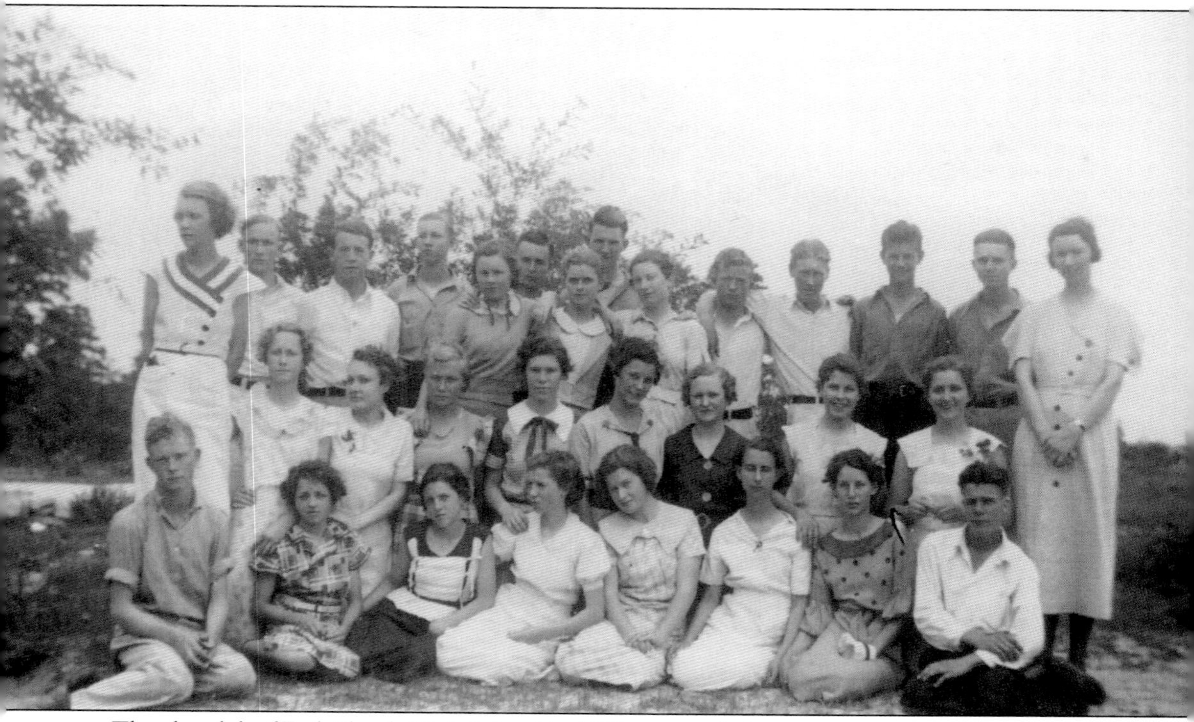

The glee club of Fork Shoals High School in 1935 consisted of nine sopranos, nine altos, six tenors, and six bass. Students were often members of multiple clubs at the school, such as the Future Farmers of America, the Leaders' Club, piano class, tennis club, the Goldsmith-Ross Literary Society, volleyball teams, and the girls' and boys' basketball teams. (Courtesy of Edwin Terry.)

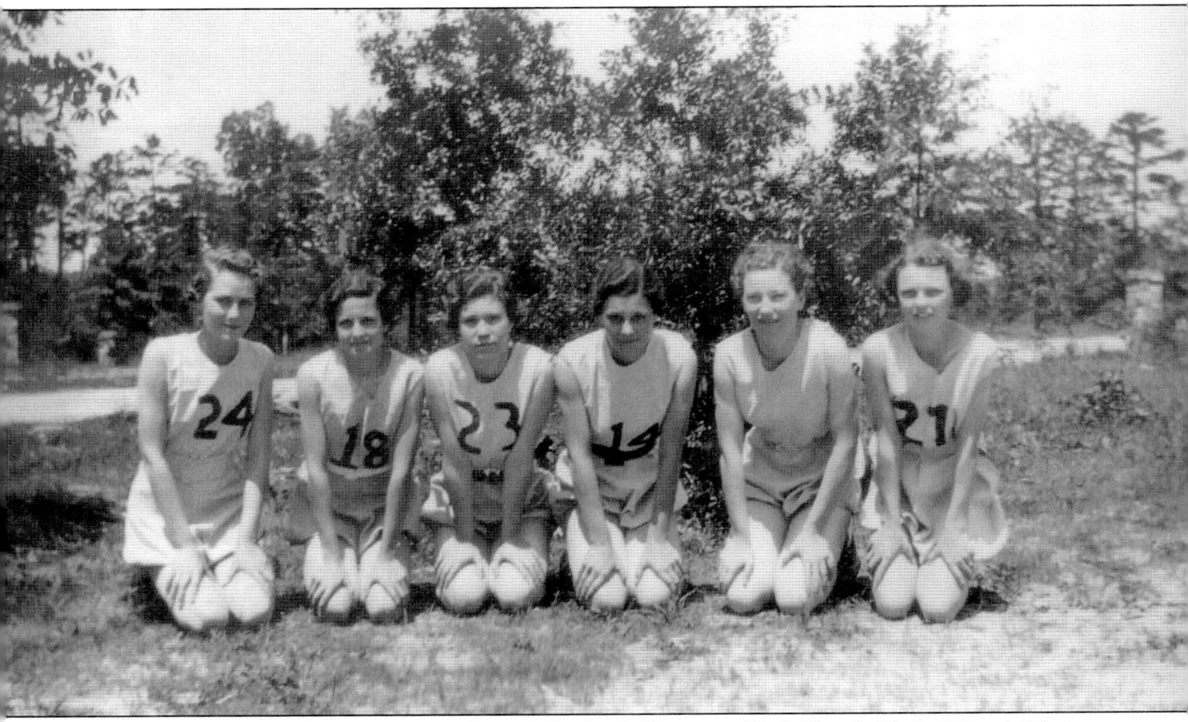

In 1935, there were eight members of the girls' basketball team at Fork Shoals High School. Their coach was W.A. Owings. The captain of the team was Doris Turner Crawford. The team won four games against other area high schools that year. Doris Turner Crawford was also a member of the tennis club and the glee club. (Courtesy of Edwin Terry.)

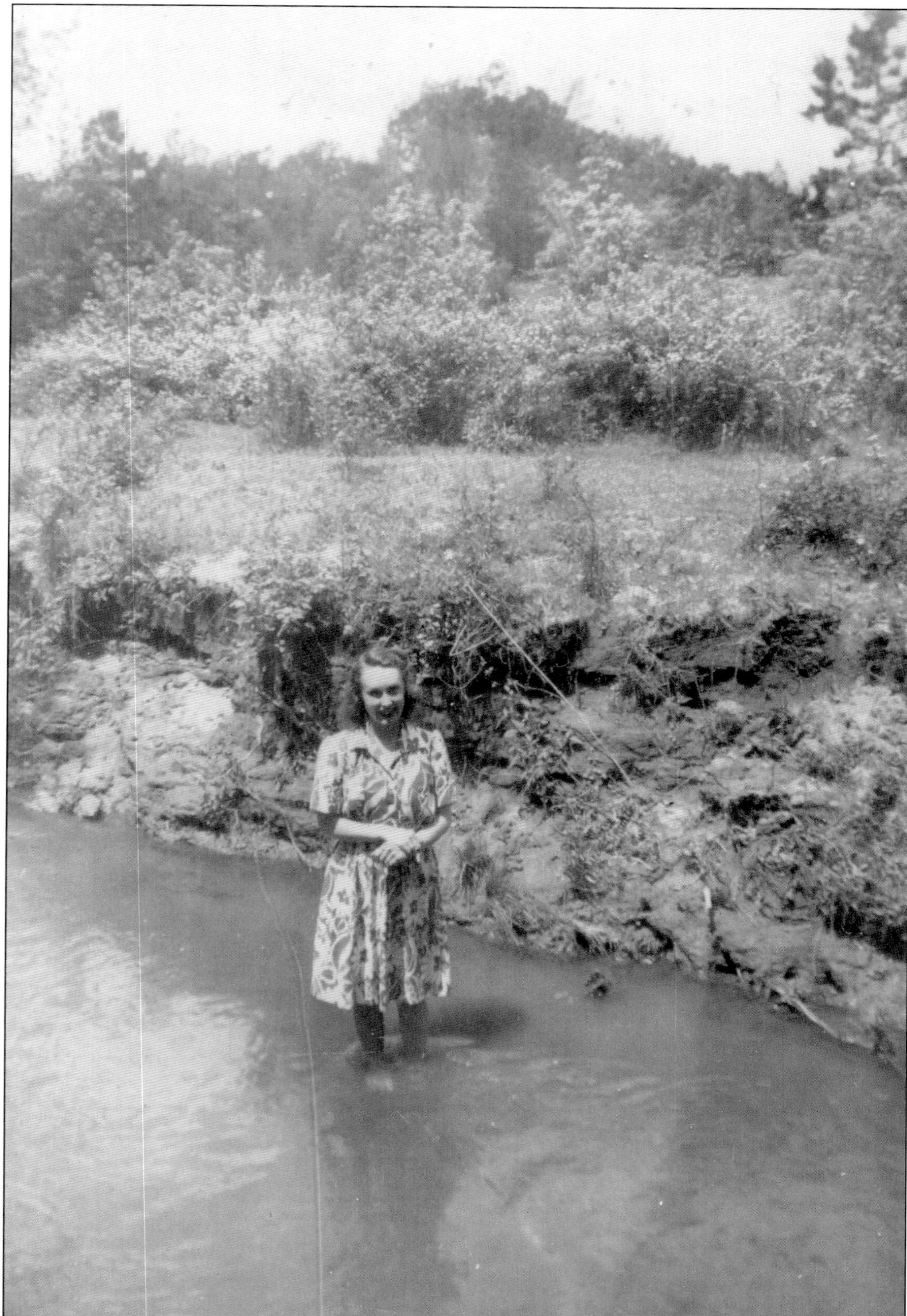

Wilma Rodgers Gault taught science and history for many years. She is seen here wading in one of the area creeks. (Courtesy of the Gault family.)

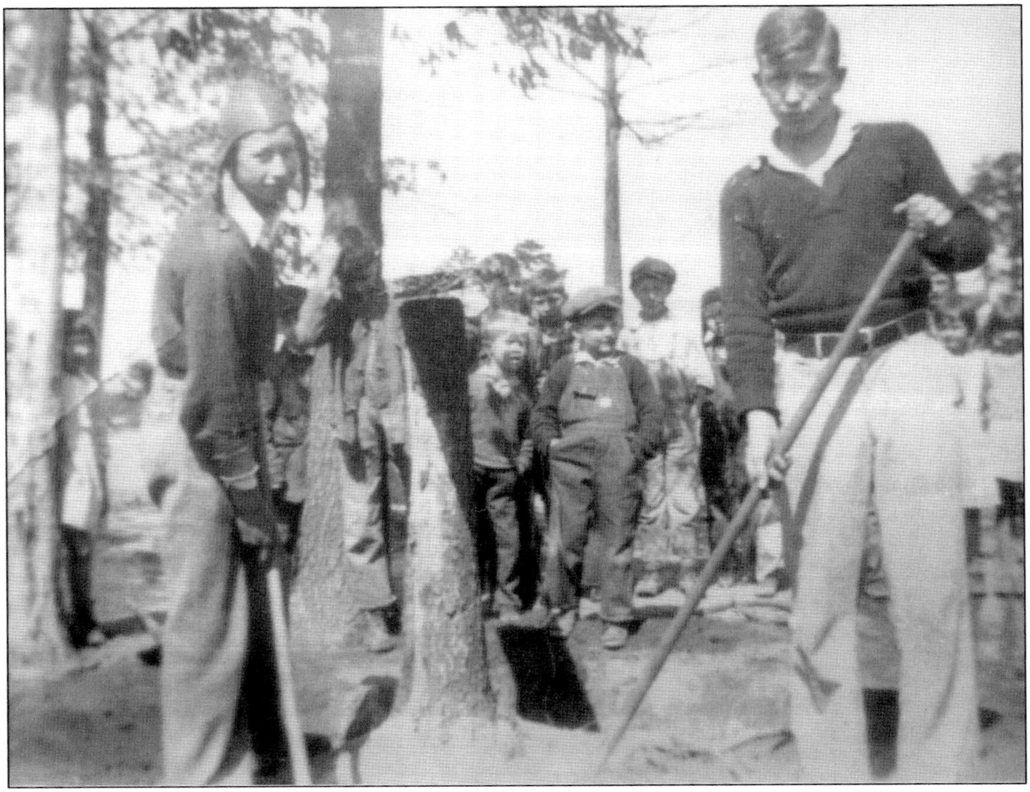

In 1936, Fork Shoals students chop trees to clear the land for a new building. These were the Future Farmers of America and other students who participated in almost all community events. The students were very involved in extracurricular activities, academics, and sports. (Courtesy of Cheryl MacKnight.)

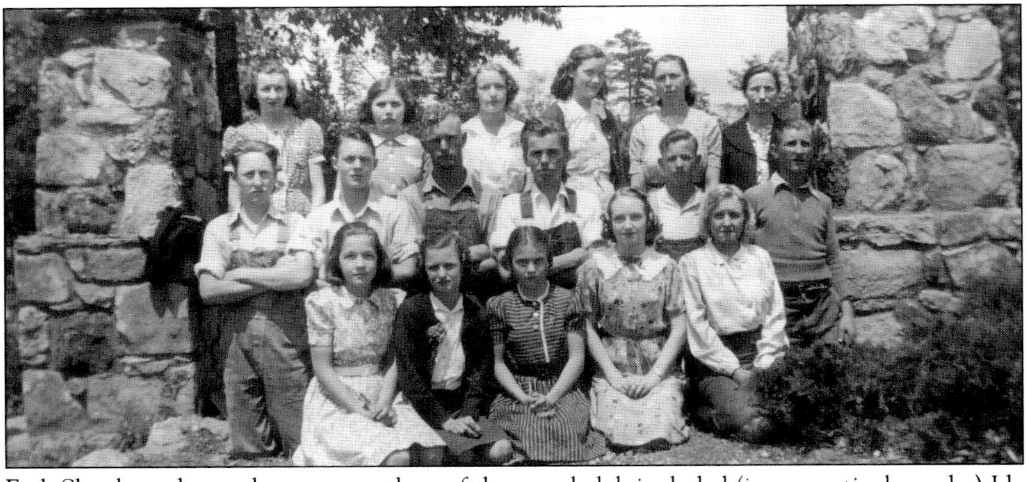

Fork Shoals students who were members of the travel club included (in no particular order) Ida Craddock, Olive Sims, Frank Neves, Nella Ruth Tumbling, Thelma Turner, Evelyn Moore, Harley Woods, Horace Kellett, Kathleen King, Swanie Storay, Anita Kellett, Leo Thomason, Peggy Sims Smith (pictured front center), Lavinia Sullivan, Doris Sullivan, Louise Campbell, Charles A. Thompson, Clayton Kirby, Zelpha Hughey, Boyd Pressley, and Ansel Gambrell. (Courtesy of Cheryl MacKnight.)

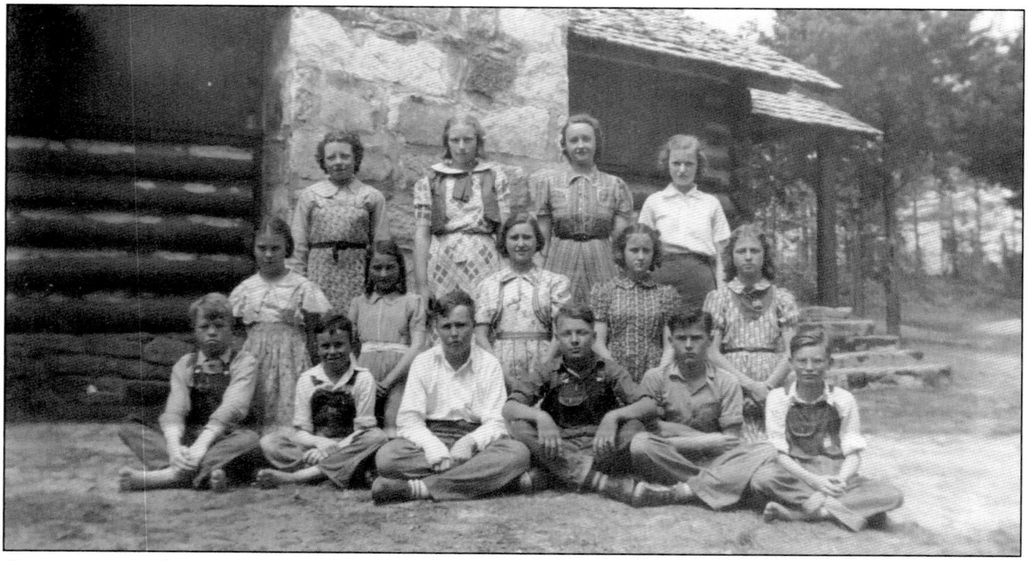

Seventeen students made up the seventh-grade class in 1939. Peggy Sims Smith (kneeling on left) was president. The seventh-grade students that year were (in no particular order) Vera Mae Abercrombie, Julia Annie Sims, Buster Kellett, Peggy Sims (Smith), Horace Kellett, Willie Jane Sims, Mary Ruth Mattison, Hazel Smith, James Meeks, Ray Talley, Virginia Meeks, Dorothy Terry, Mellmoth Riddle, Marion Tripp, James Robertson, Nella Ruth Tumblin, and J.C. Williams. (Courtesy of Cheryl MacKnight.)

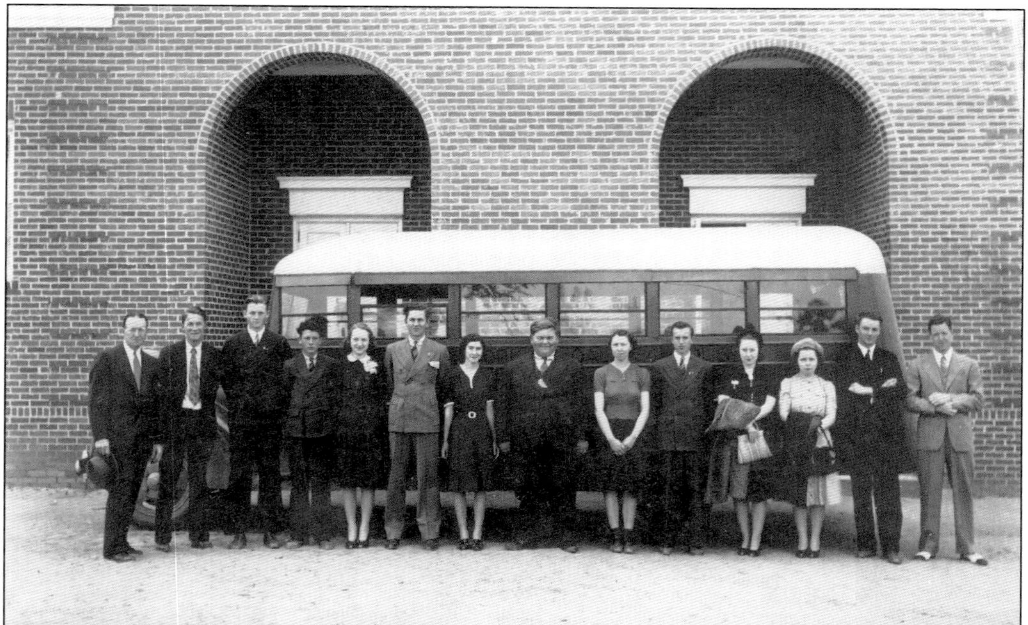

Fork Shoals graduates and faculty of 1940 posed in front of their new auditorium. They were (from left to right) superintendent W.E. Sims and seniors Ansel Moore, Carlos Kellett, Bill Scott, Ida Craddock, Alton Henderson, Frances Rodgers, Ulyss Childress, Kathleen King, Julian Sullivan, teachers Amelia Beason (Woods), Sara Belle Brooks, John Smith, and principal R.M. Stoddard. Graduates not shown were Lois Cox, Dick Griffin, Walter Smith, and Doris Sullivan. (Courtesy of Cheryl MacKnight.)

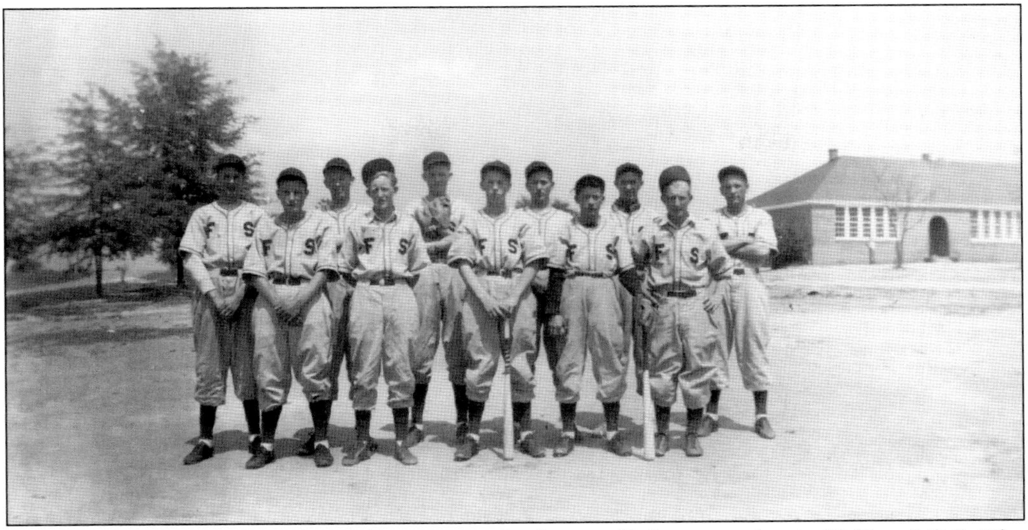

The Fork Shoals High boys' baseball team won the district 10 championship in 1940 and 1941. The 1940 team, with the grammar school building in the background, consisted of (in no particular order) Carlos Kellett, Alton Henderson, Bill Farrow, Bill Scott, Bruce Farrow, Swanie Storay, Dick Griffith, Wilbur Mahon, Warren Sprouse, Wayman Lollis, Leo Thomason, and Harley Woods. (Courtesy of Cheryl MacKnight.)

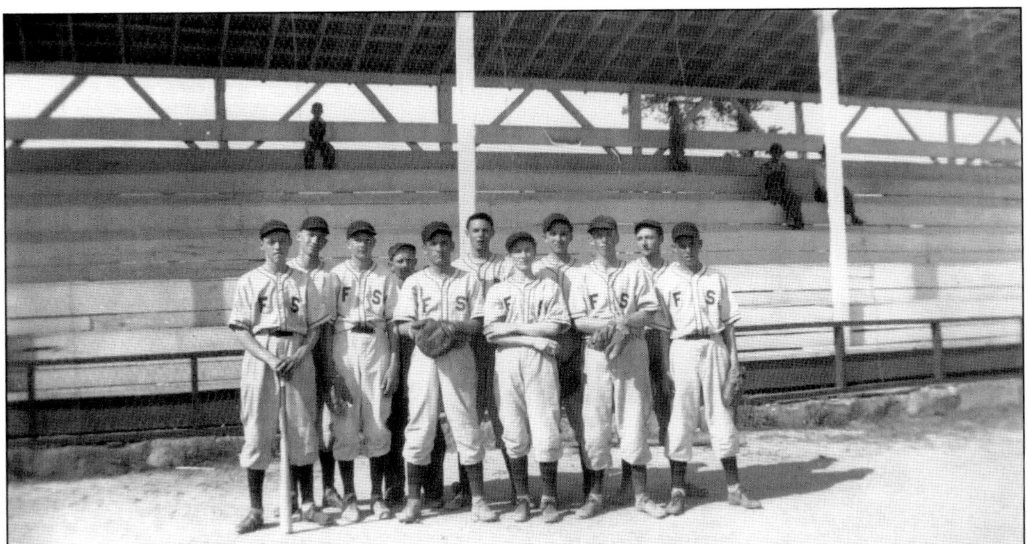

The 1941 team consisted of (in no particular order) Bill Farrow, Warren Sprouse, Bruce Farrow, Johnny Jenkins, Ed Ridgeway, Duard Rodgers, Calvin Sullivan, Willard Mchaffey, Wayman Lollis, Howard Smith, Harley Woods, Harry Farrow, Carl Tumblin, Billy Sims, Leo Thomason, and Swanie Storay, coached by John Smith. This team is shown at the Lebanon Stadium, a ballpark built by William David Ridgeway. (Courtesy of Cheryl MacKnight.)

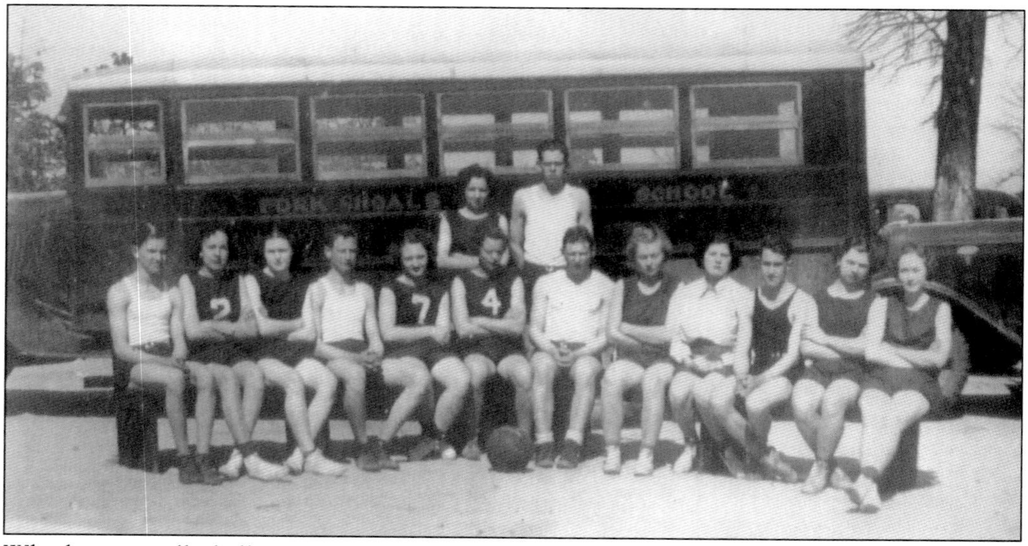

Whether to a volleyball game in 1940, or another trip to Washington, DC, the school made good use of these buses. These long-bodied T-Model buses had benches running the length of them. The aisle was often full of standees. There was a shovel on board for digging out of muddy spots in the roads. After the Great Depression, the school was fortunate to have these buses. (Courtesy of Cheryl MacKnight.)

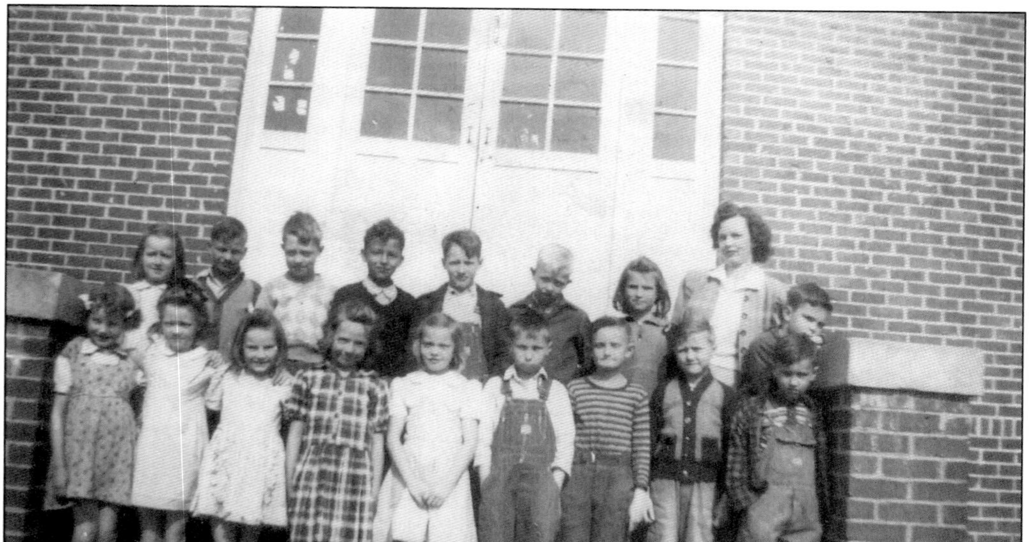

Katherine Ross, photographed with the second- and third-graders of Fork Shoals School in 1944, taught second grade for 30 years. She worked in the school library for 10 more years and retired in 1980. A Fork Shoals graduate of 1939, she taught in Belton for four years and then returned to her alma mater. Her father was Dr. W.A. Ross. (Courtesy of Scipio Peden.)

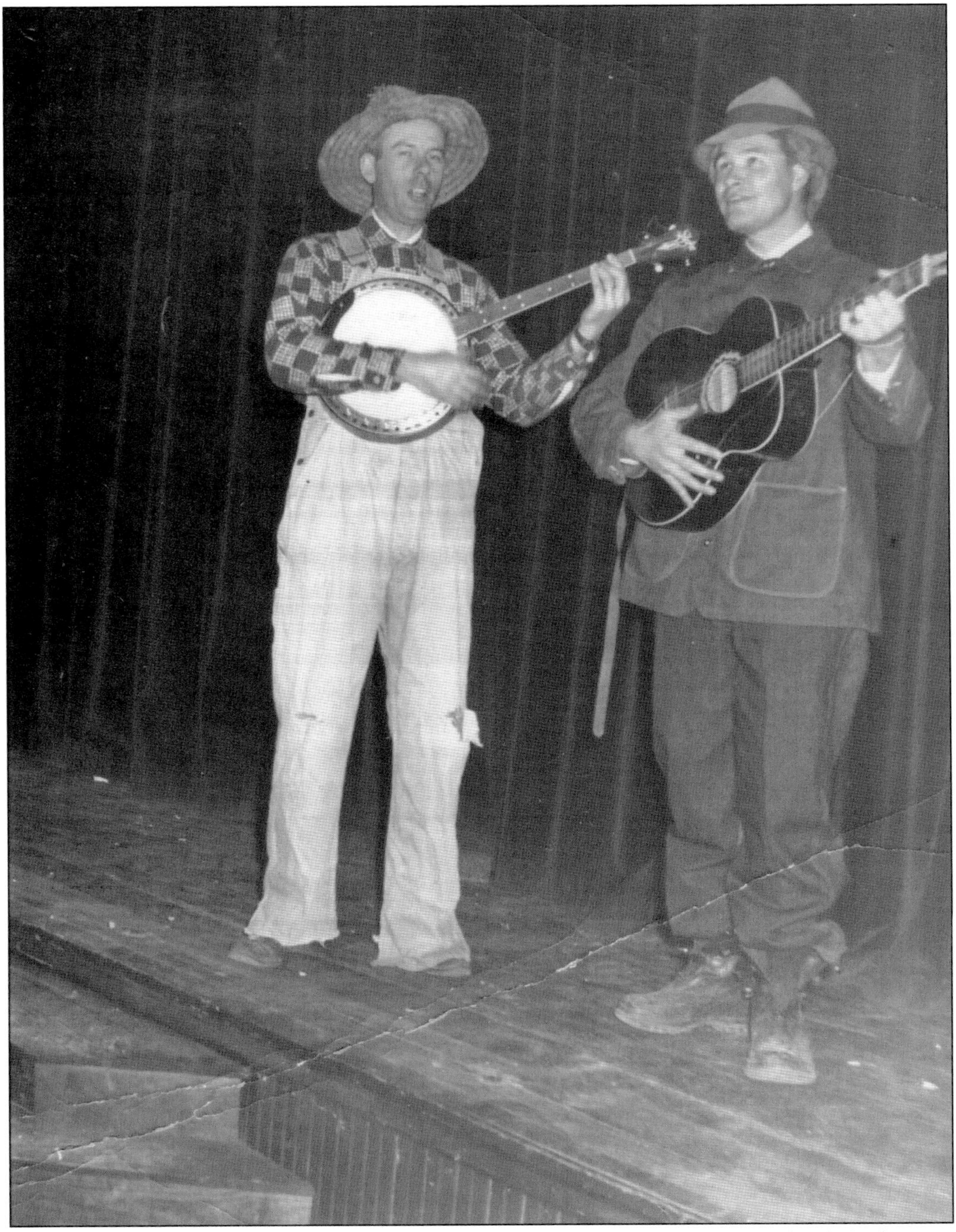

The Fiddlers Convention was a huge annual gathering of talent at Fork Shoals School. Orchestras, harps, horns, singers, entertainers, and musicians competed for cash prizes. Nearly everyone participated, as did Supt. G. LaFoy Woods (left) and agriculture teacher Richard Cothran (right) in 1952 on the old auditorium stage. Beginning in 1925, it was held each year on February 22 to celebrate George Washington's birthday and continued well into the night. The prizes ranged from 25¢ up to $8 for the first place for orchestra. Admission was 20¢ for adults and 10¢ for children, likely increasing as the years went by. The community looked forward to this event every year. (Courtesy of Cheryl MacKnight.)

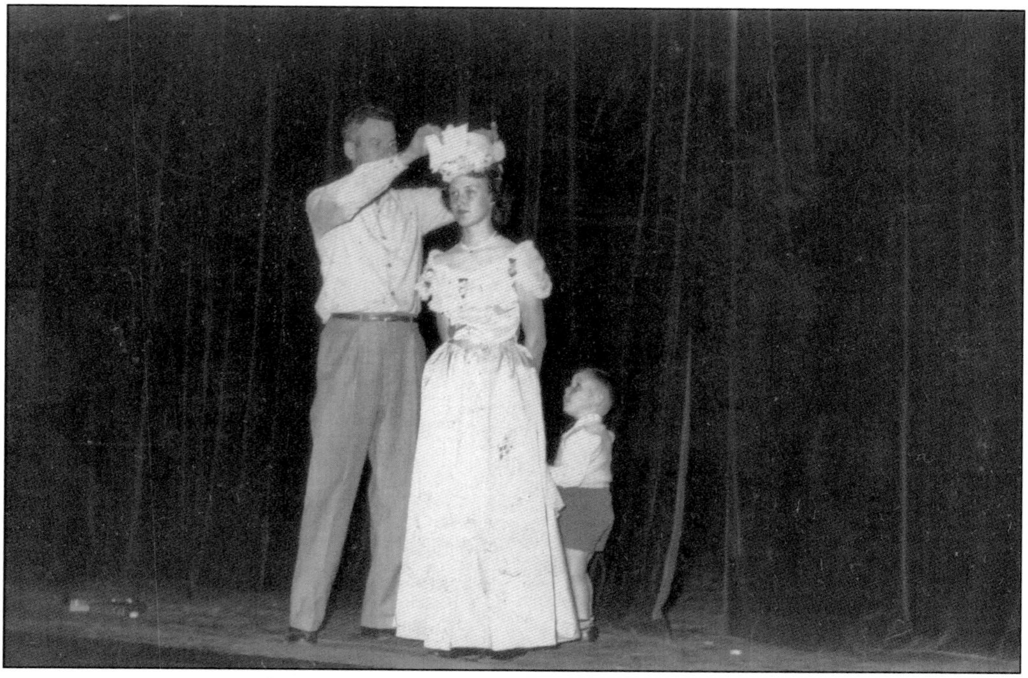

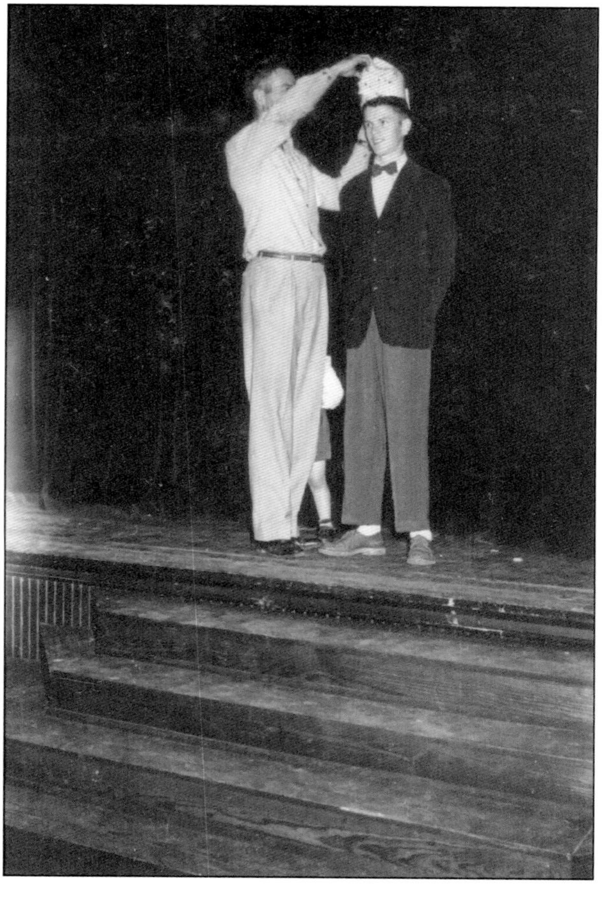

In 1951, Supt. G. LaFoy Woods crowned Joyce Kellett (Massey) queen, as his son Woody looked on (above). Charles Heatherly was crowned king that year (left). Woody was still on stage with his dad. An annual competition at the Halloween carnival, these two students earned the most pennies per vote for the title of queen and king that year. (Both, courtesy of Cheryl MacKnight.)

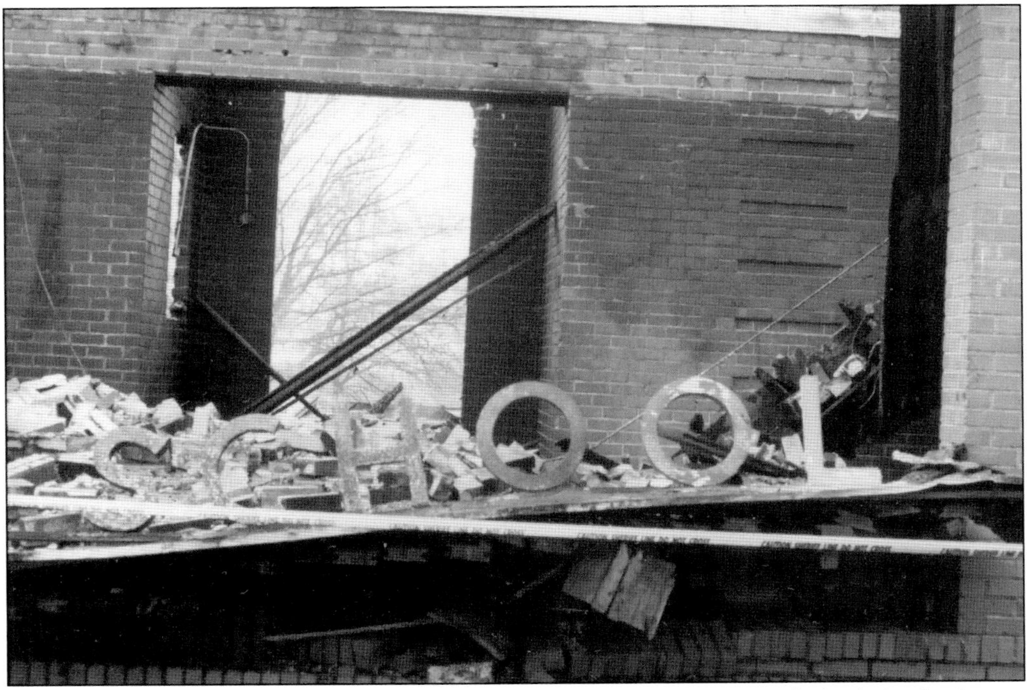

Fork Shoals School lost its auditorium to fire in April 1991. It was originally built for the former high school in 1939 and was located in-between the grammar and high school buildings. The high school building close by suffered some damage but was repaired. (Courtesy of Lil Mitchell.)

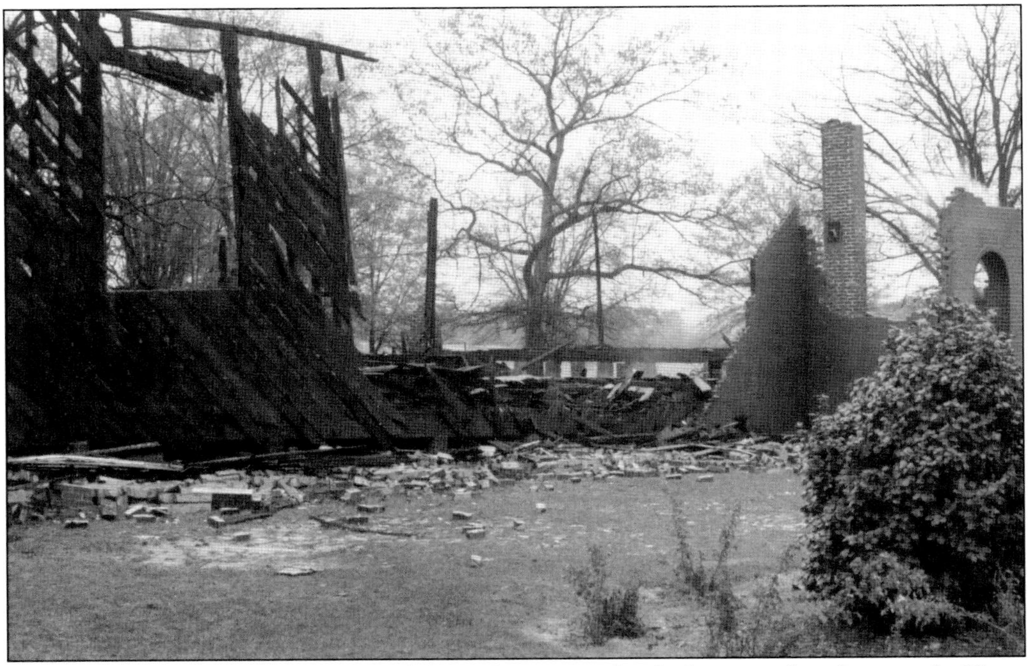

The intensity of the fire that destroyed the auditorium is clearly shown in this photograph. There had been a number of brush and house fires at that time, which prompted an arson investigation. Having survived setbacks in previous years, whether it was a war or fire, the school and community still thrives. (Courtesy of Lil Mitchell.)

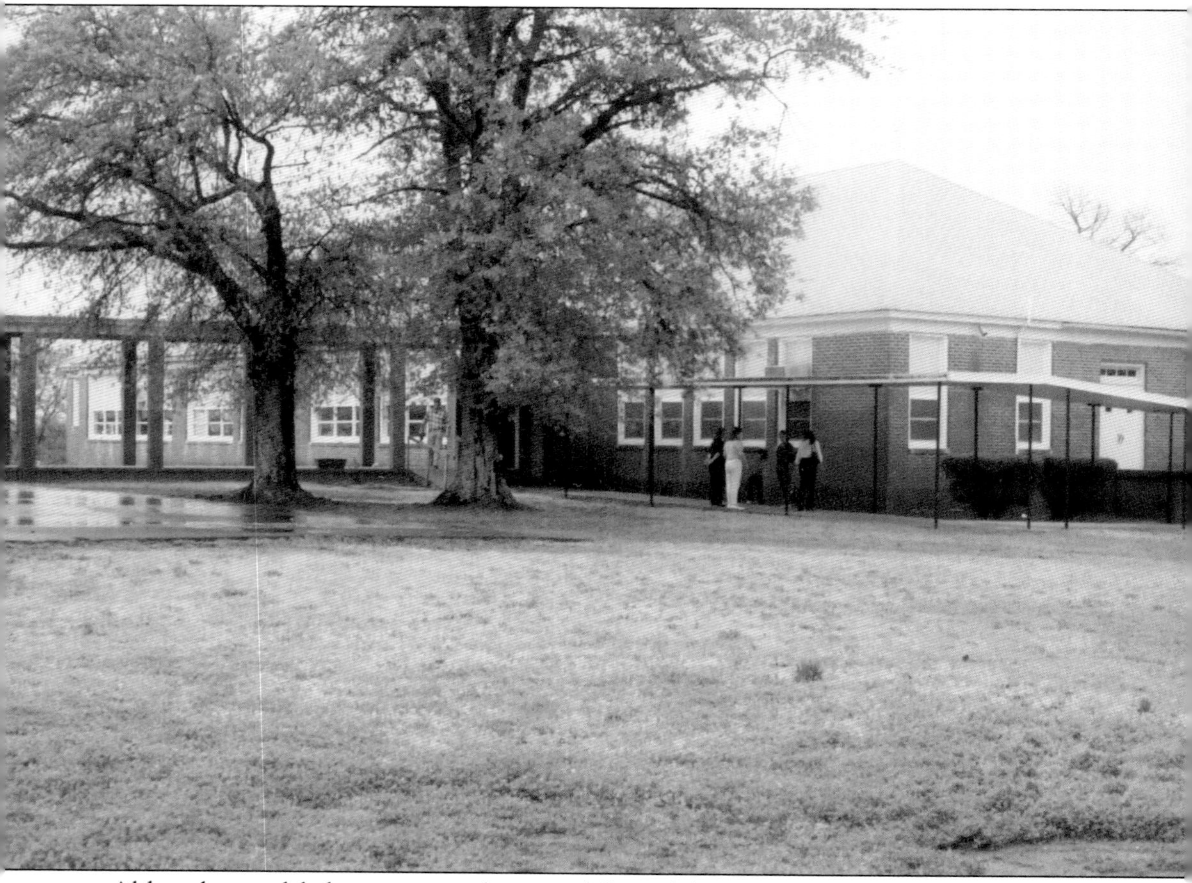

Although remodeled many times, the ground floor of the original 1916 section of the school was in use for more than 80 years. It was demolished when the new school was built in 1998, but salvageable parts were collected for reuse elsewhere. Only one of the three 1939 buildings used for the grammar grades is still in use today as a community building. (Courtesy of Lil Mitchell.)

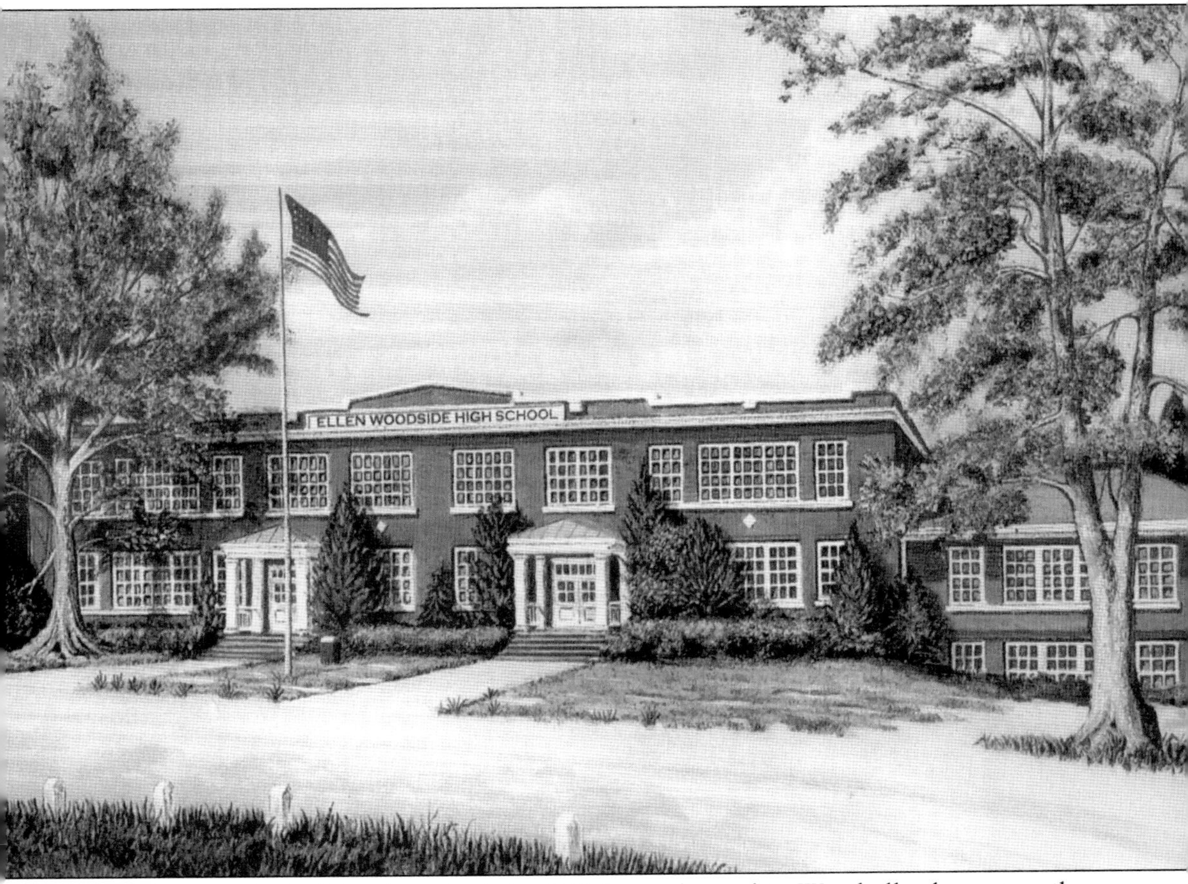

Ellen Woodside High School was built in 1925. It was located in Woodville about six miles northwest of Fork Shoals. John T. Woodside donated land and half the building cost. In the late 1930s, a large wooden community building was constructed at the north end of the school. It housed the home economics department with a kitchen, sewing room, and the agriculture and shop departments. (Courtesy of Virginia Agee Dean.)

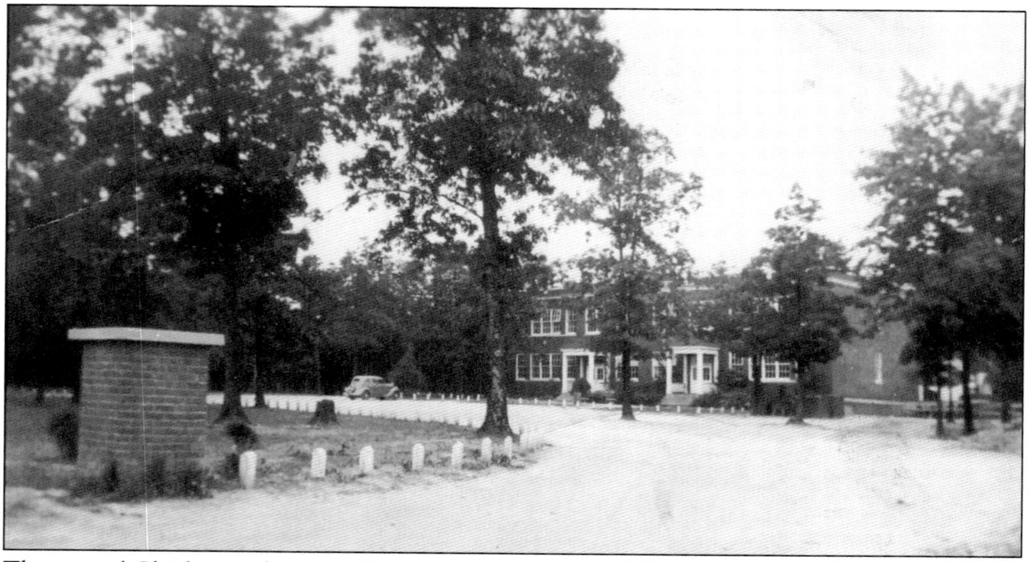

The annual Chicken and Possum Supper took place at the Ellen Woodside High School followed by a basketball game. After World War II, the school provided evening classes for a veterans' school. Until the late 1940s, a six-week summer session of school was held and the fall session did not begin until mid- to late October, when crops had been gathered in. Smaller schools consolidated to Ellen Woodside High until 1965, when it became an elementary school. (Photograph by Hattie Sims.)

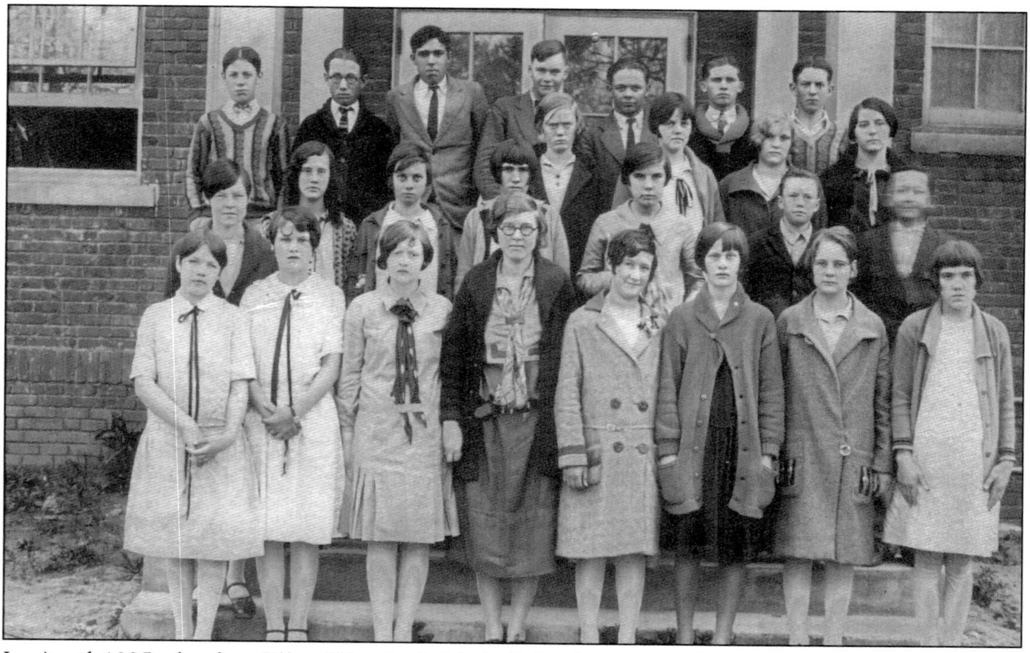

In April 1925, the first Ellen Woodside High School students finished their year in the new building, after having started the school term in the community hall over the Woodville store, with W.D. Iva as their first principal. This real-photo postcard of eighth-grade students was taken in 1926. (Courtesy of the Ware family and Linda Hufstetler.)

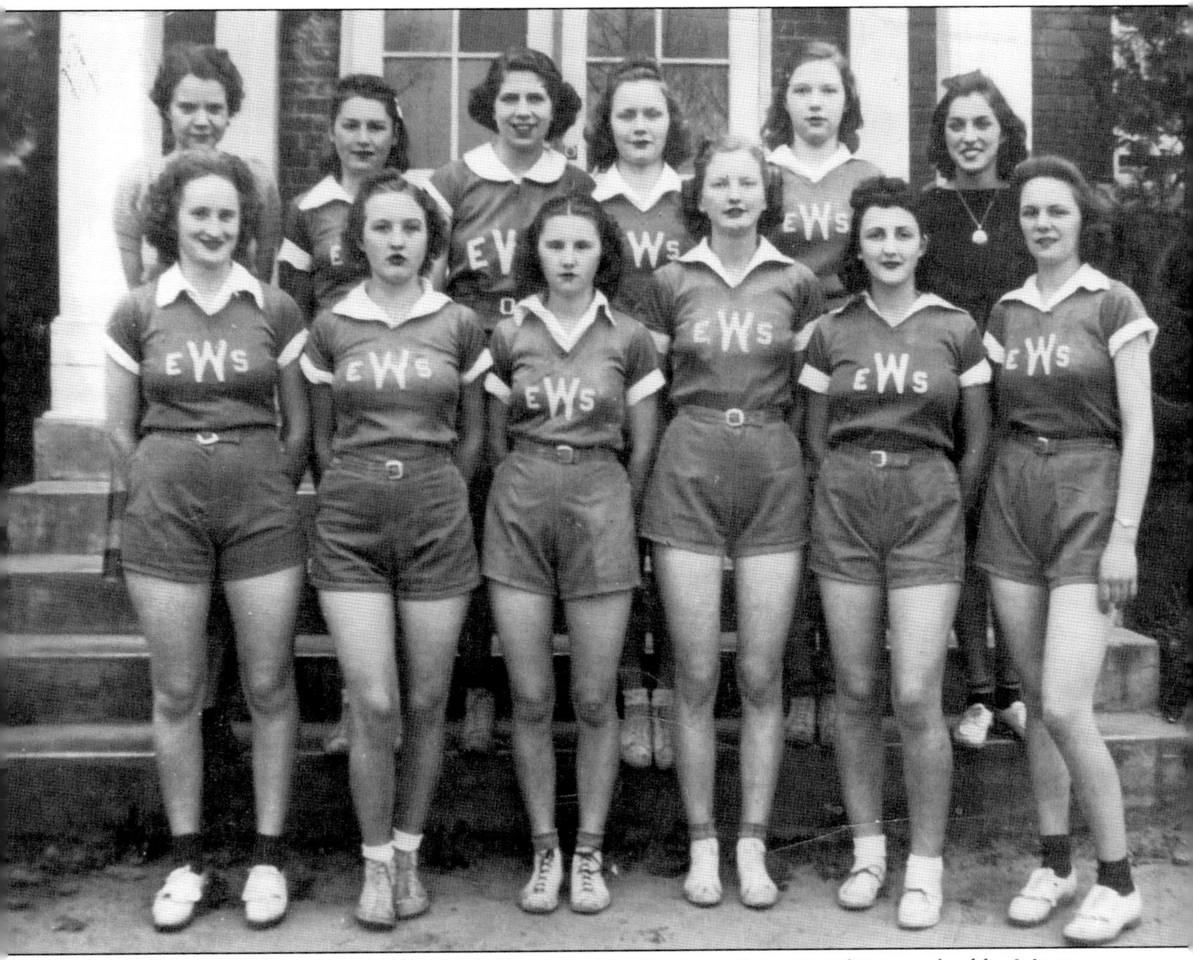

The Ellen Woodside 1938 girls' basketball team won the state championship coached by Minier Padget. The team consisted of, from left to right, (first row) Elizabeth Chapman, Frances Chastain, Louise Gilliam, Mary Huff (Yardbray), Dot Harris (Stroud), and Frances Gilliam (Coker); (second row) Mary Louise Humphries, captain Frances Darby (Hillhouse), Ruth George, Jean Hammond (Coogler), Nancy Charles, and manager Helen Babb (Whitney). (Courtesy of the Hillhouse family and Linda Hufstetler.)

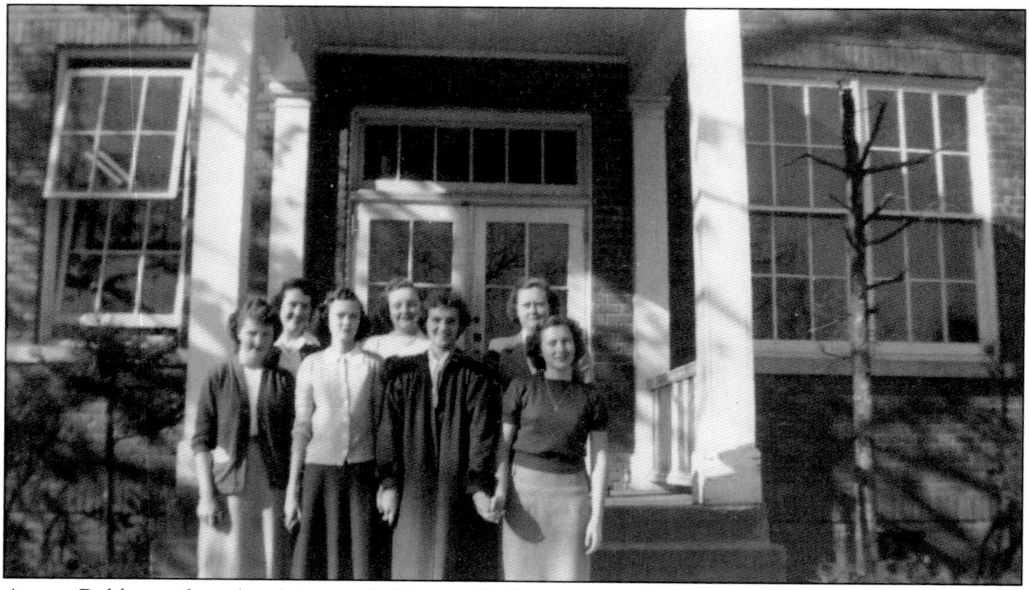

Agnes Babb taught school in south Greenville for many years. She instructed these senior girls in shorthand, typing, and bookkeeping in 1951 at Ellen Woodside High School. Shown here are, from left to right, senior treasurer Helen Campbell (Coker), senior secretary Wylda Chandler, senior vice president Ethel Alverson (Stubbs), Charlotte Jordan (Cartee), visitor Virginia Alberson (Hilley), Agnes Babb, and Doris Allen (Taylor). (Courtesy of Virginia Agee Dean.)

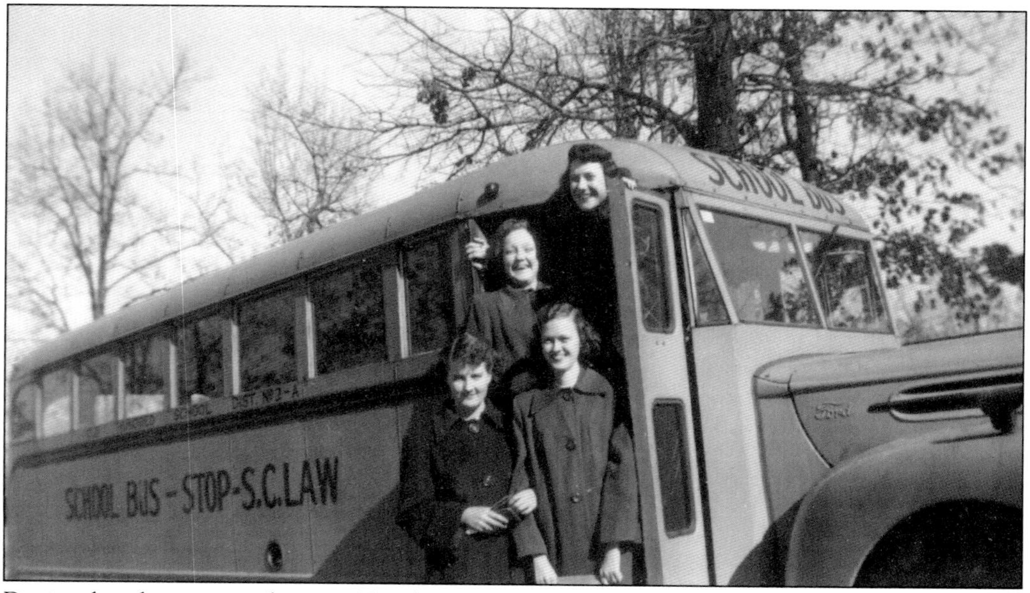

During lunchtime one day in 1951, these girls struck a pose in the door of the bus at Ellen Woodside High. They were to be graduates of the high school that year. They are (top right) Wylda Chandler, (middle) Charlotte Jordan (Cartee), (bottom left) Helen Campbell (Coker), (bottom right) Ethel Alverson (Stubbs). (Courtesy of Virginia Agee Dean.)

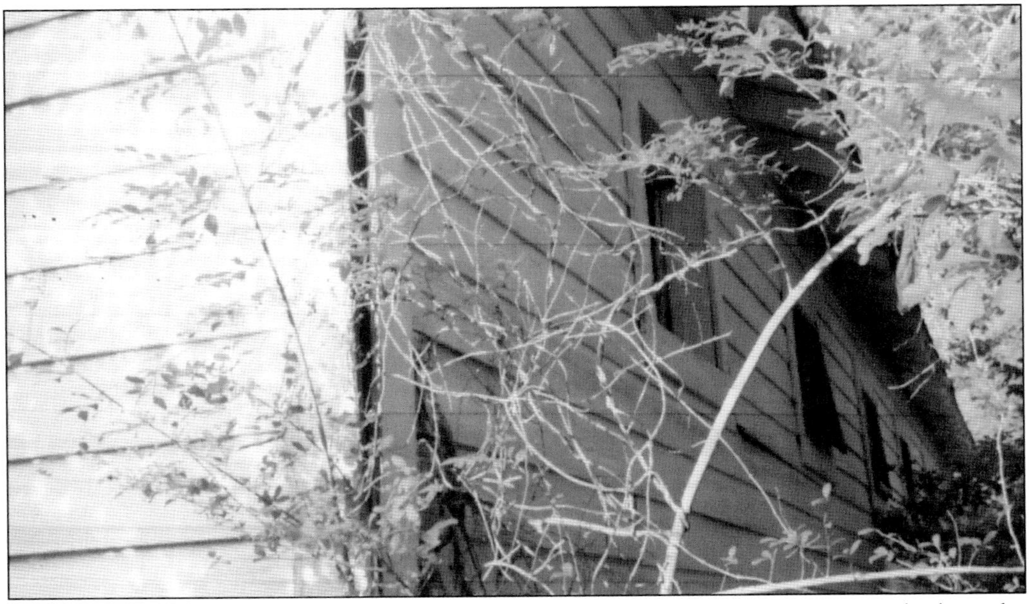

Chapman Grove School was a school for blacks in the Fork Shoals district. It was built in the early 1920s and named for Prof. John H. Chapman. Teachers boarded in his family's residence. The school had multiple buildings, one each for elementary, middle, and high school grades; a large cannery (above); potato shed; laundry; and a mechanic training shop. It closed in 1954. (Photograph by Cheryl MacKnight.)

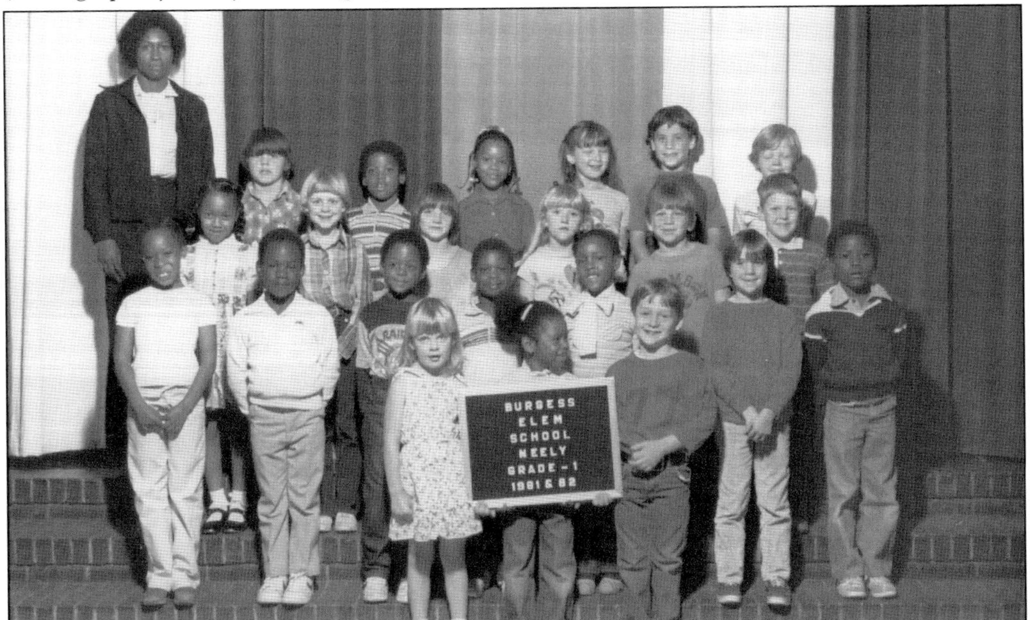

County schools were integrated in 1970–1971. Burgess School and Ellen Woodside were organized as companion schools. Primary grades attended Burgess, and third through seventh grades attended Ellen Woodside. The arrangement preserved the black community's school (Burgess). Gloria Neely and her first-grade class were photographed, and the photograph as well as news clippings and other pictures of the school year, were presented to Principal Peggy Sims Smith. (Courtesy of Cheryl MacKnight.)

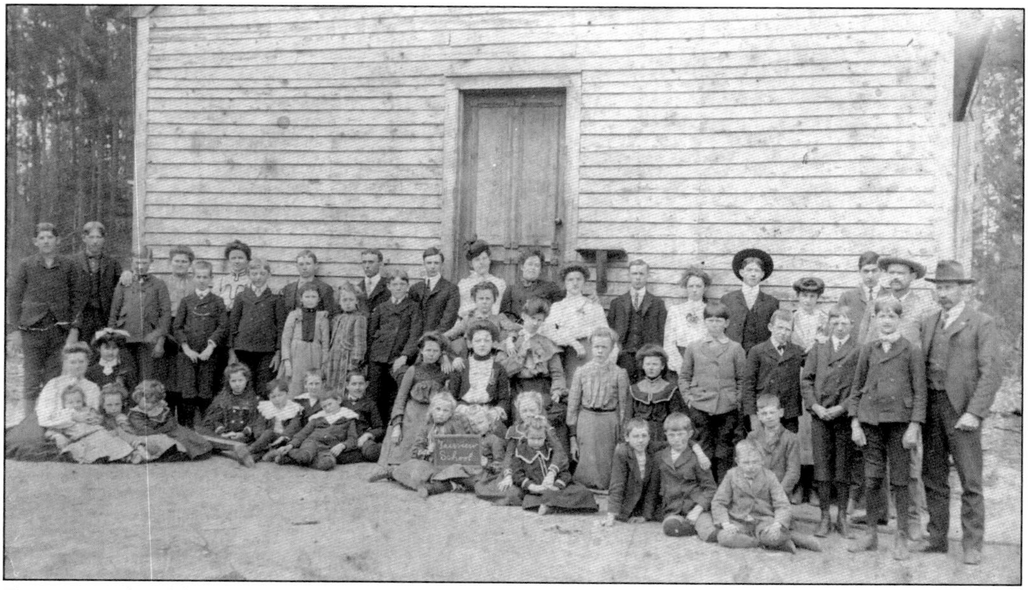

Fairview School held classes into the 1900s. This area east of the Reedy River was settled in the late 1700s, predominately by Scotch-Irish immigrants. Well into the 1800s, church buildings, many of them log structures, were often used as classrooms for the children of the community. As the area became more populated, school buildings such as this one were built. (Courtesy of Edwin Terry.)

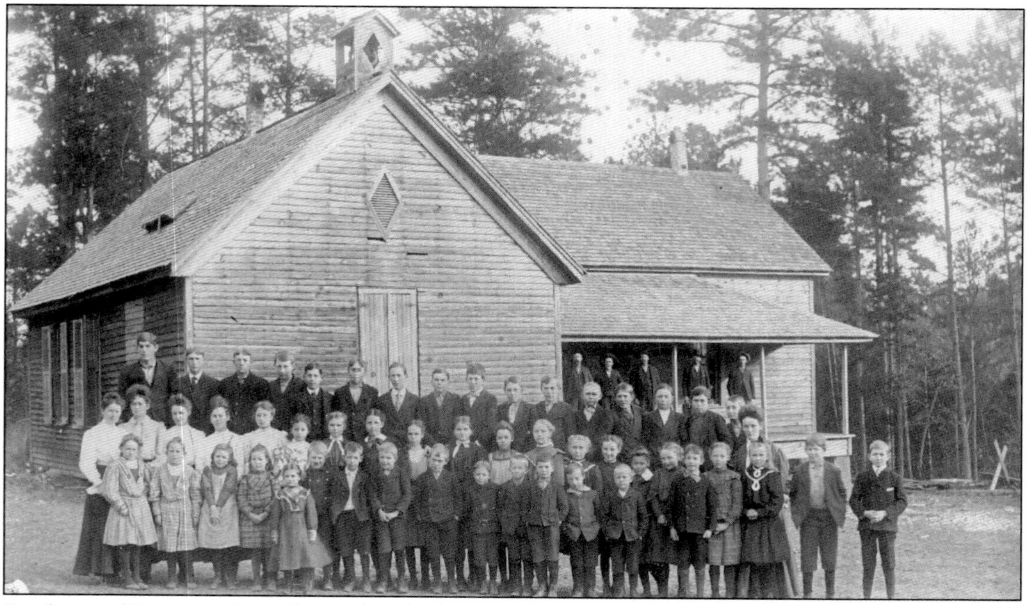

Students of Fairview School posed with their teacher, Nannie Babb (second row, far left), for this photograph sometime during the early 1900s. The school was located near Fairview Presbyterian Church, which was established in 1786, at the location of a cold spring about two miles east of the Reedy River in southern Greenville County. The school building is no longer standing. (Courtesy of Edwin Terry.)

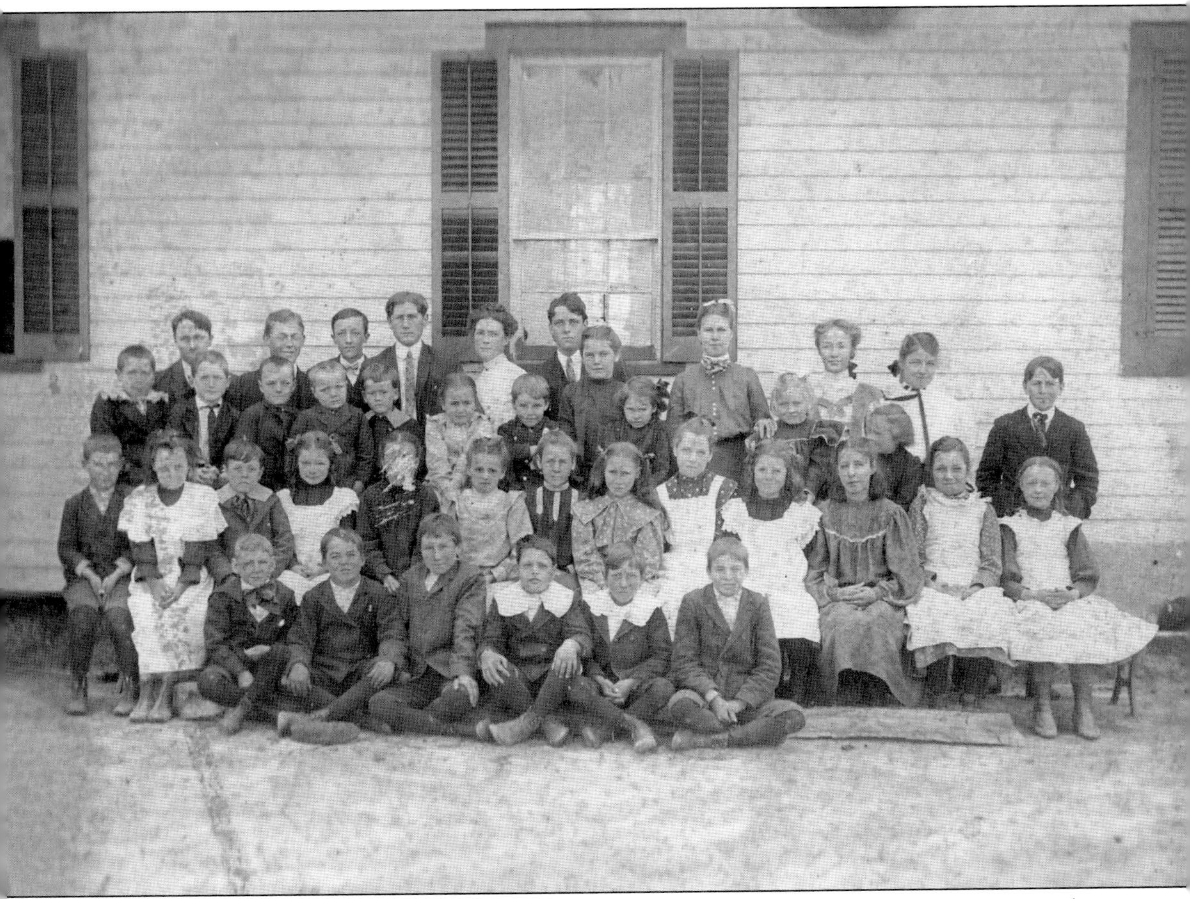

Chandler School was located along the eastern side of Augusta Road in Greenville County, south of Daventon Road. It served the children in the Chandler community from about 1900 until the early 1950s. After the school closed, the building was used as a community social facility. (Courtesy of Nancee Lee Knight Yearick.)

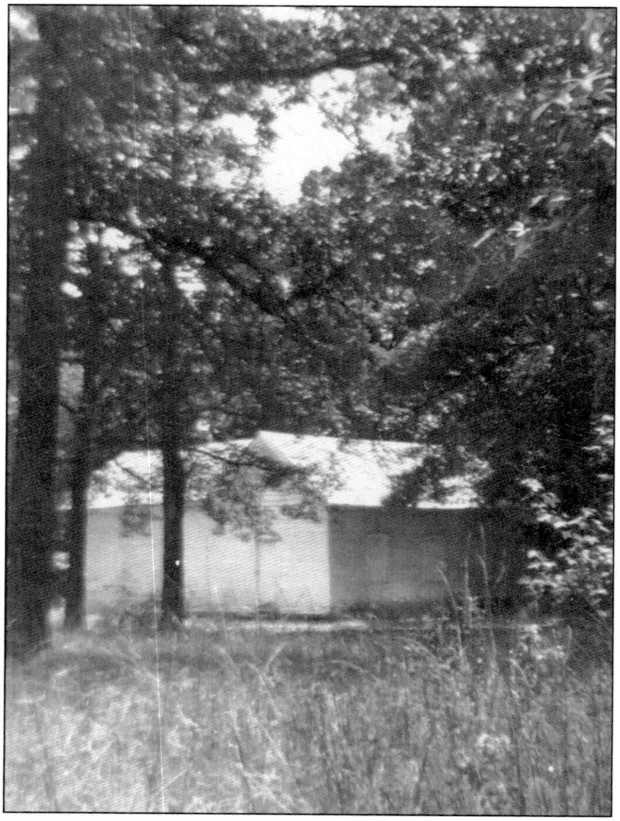

Rev. C.L. Stewart preached in a small schoolhouse on Ellen C. Woodside's property when traveling by horse between Piedmont and Fairview. Because of her diligence, in 1882, the Lickville Presbyterian Church was built on a site provided by W.A. McKelvey. Its session house became Lickville School. The student body moved to McKittrick Road in the 1920s, but it closed in the early 1950s. Both buildings are still standing. (Courtesy of Lickville Presbyterian Church)

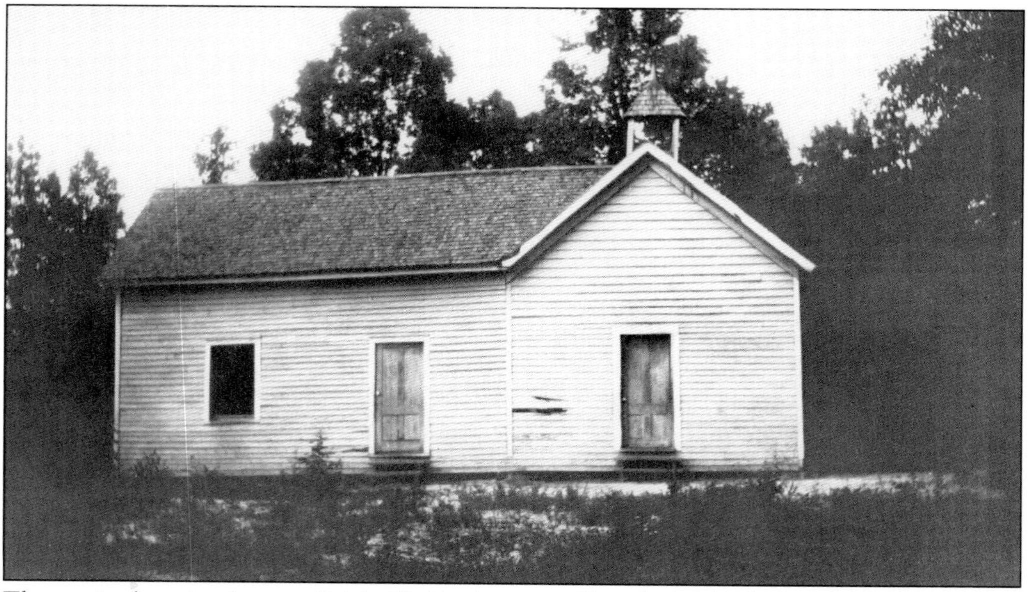

The original session house of Lickville Presbyterian Church, established in 1882, was used for Lickville School. It was a strong school in its day and continued classes at this location until the 1920s. In 1968, rooms were added, and it was made into an education building of the church. It is still in use today as a social center. (Courtesy of James H. Woodside, Woody Woods, and Lickville Presbyterian Church.)

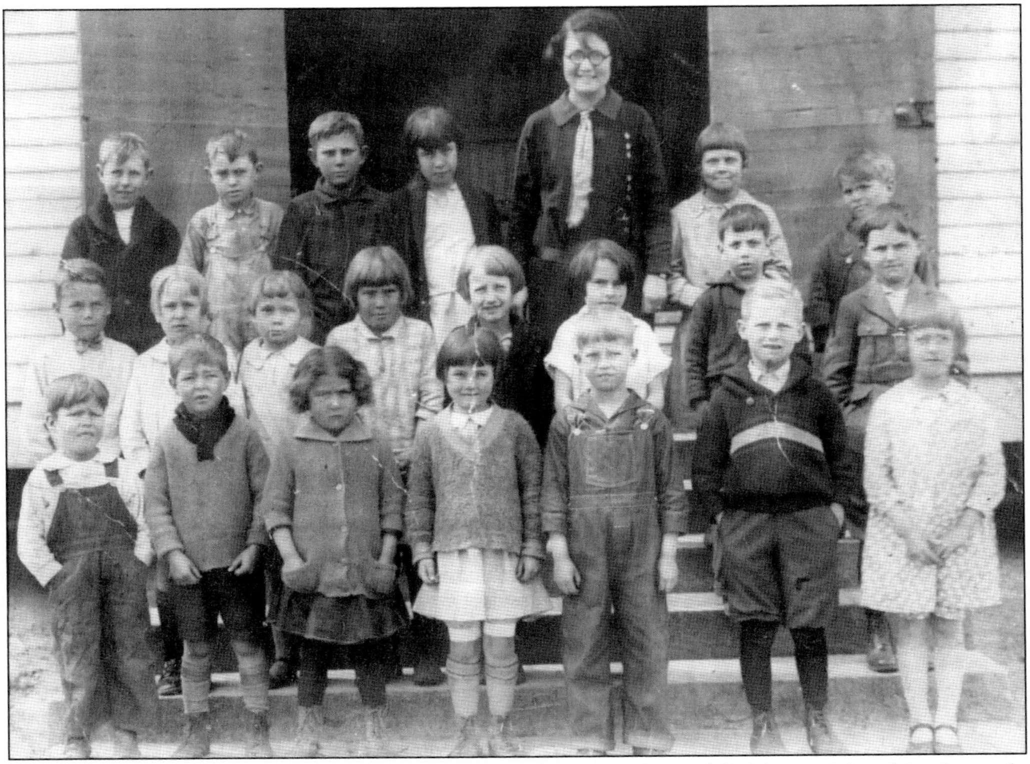

Posing their best are students and Mrs. A.W. Hawkins, teacher, of Oaklawn School in the early 1900s (above). The school was on a hill at the Ware Place, located across the road and just south of the old Ware home and the intersection of Augusta and Cooley Bridge Roads. Oaklawn School was a one-teacher school, otherwise known as Oaklawn 4H. Hawkins's maiden name was Martha Mae Turner. She was born in 1905 and married Arthur Waitsel Hawkins in 1925. He was superintendent of Ellen Woodside High School, located less than two miles north of Oaklawn School. The photograph of Oaklawn School (below) was taken in 1939. (Above, courtesy of Cheryl MacKnight; below, courtesy of the Ware family and Linda Hufstetler.)

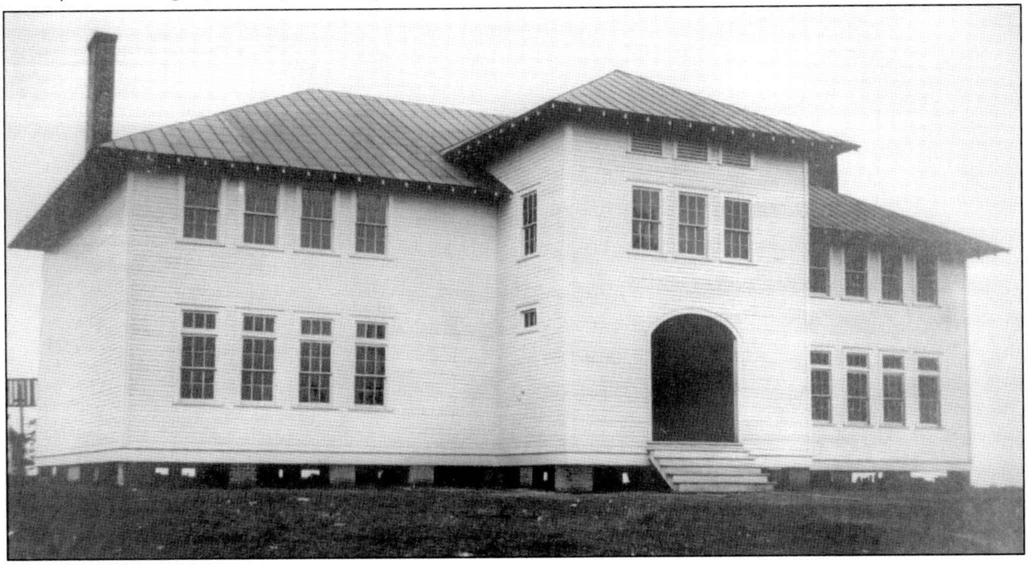

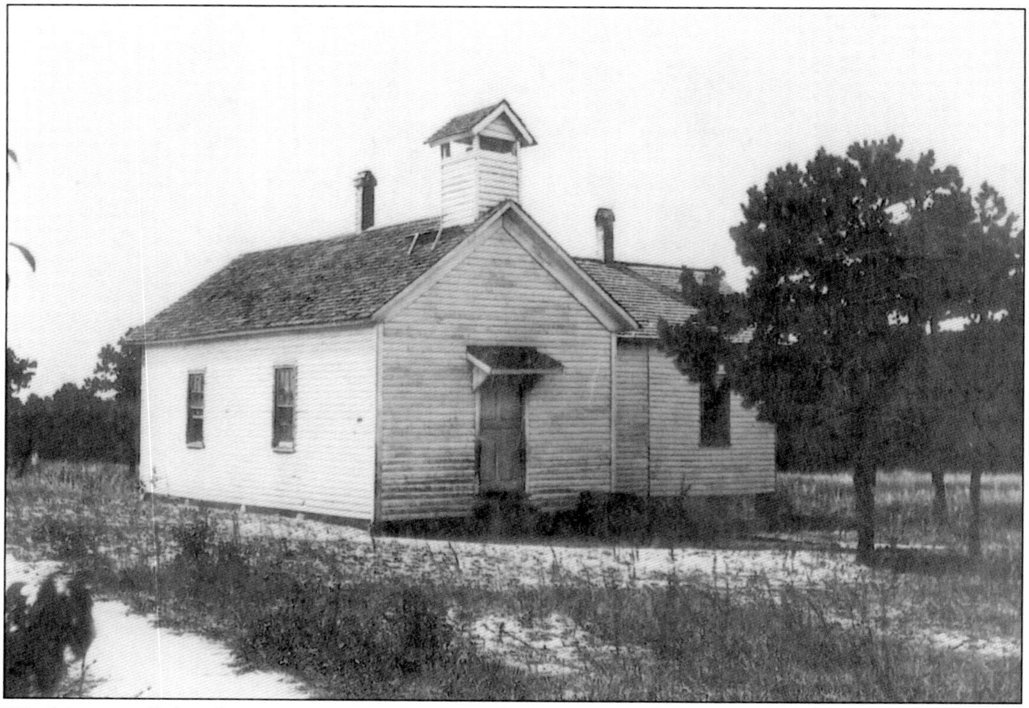

Washington School is representative of nearby historic schools not shown in this book. As the population grew, these small schools were consolidated into larger buildings. They were of familiar names, such as Columbia, Dry Oak, East View, Flat Rock, Forksville, Grove, Holly Grove, Horse Creek, Lebanon, McCullough, Old Hundred, Pine Hill, Saint Albans, Sanoma, Shady Grove, and West Dunklin. Any omissions are unintentional. (Courtesy of Joyce Coates and Linda Hufstetler.)

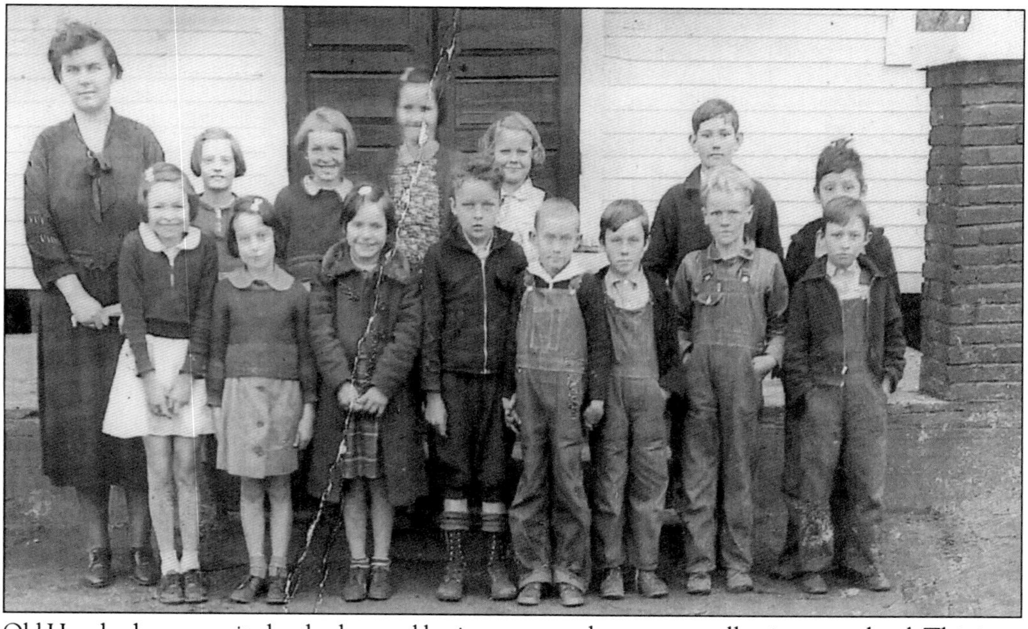

Old Hundred community has had several businesses over the years as well as its own school. This is an early-1900s photograph of an elementary class at the front of the school. (Courtesy of Jim Scott.)

Four

DYNAMIC WATERS

MILLS

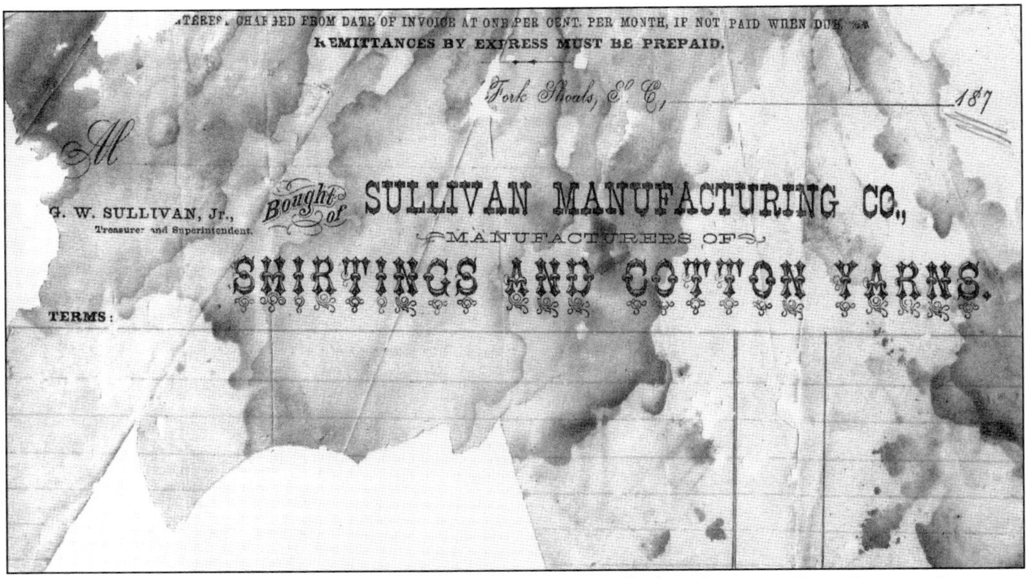
INTEREST CHARGED FROM DATE OF INVOICE AT ONE PER CENT. PER MONTH, IF NOT PAID WHEN DUE.
REMITTANCES BY EXPRESS MUST BE PREPAID.

Fork Shoals, S. C., 187

M

G. W. SULLIVAN, Jr.,
Treasurer and Superintendent.

Bought of SULLIVAN MANUFACTURING CO.,
MANUFACTURERS OF
SHIRTINGS AND COTTON YARNS.

TERMS:

The invoice shown here is one small indication of the unusually long history of textile manufacturing in Fork Shoals. Coming from the latter part of the 19th century, this document is an early glimpse of an industry's history that began in the first decades of that century. This history, surprisingly, parallels that of better-known parts of Greenville. (Courtesy of Jack Sullivan.)

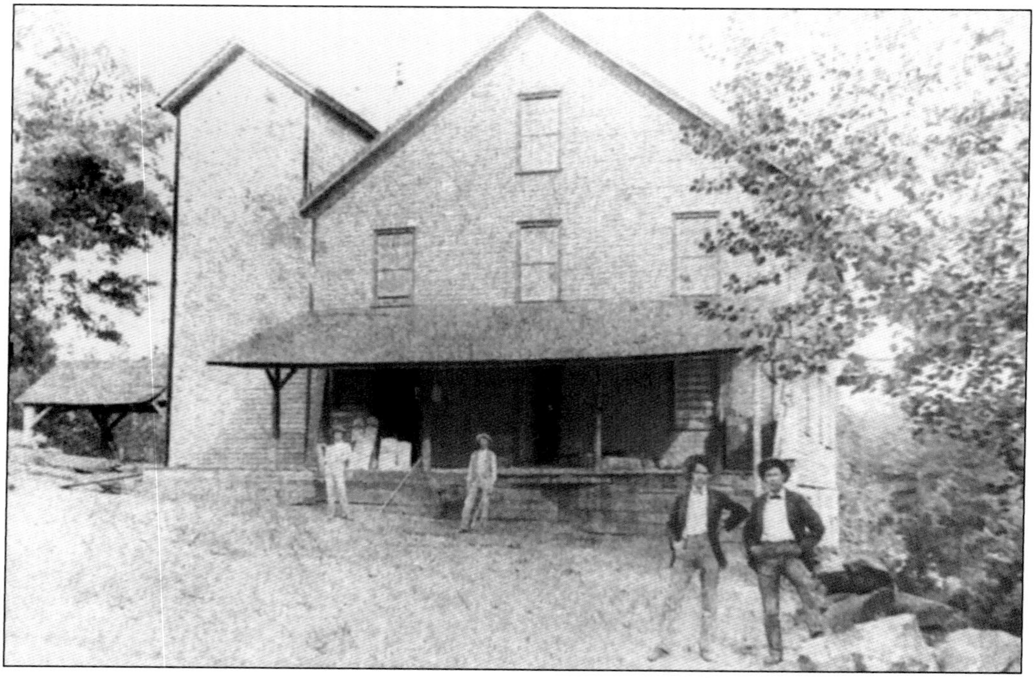

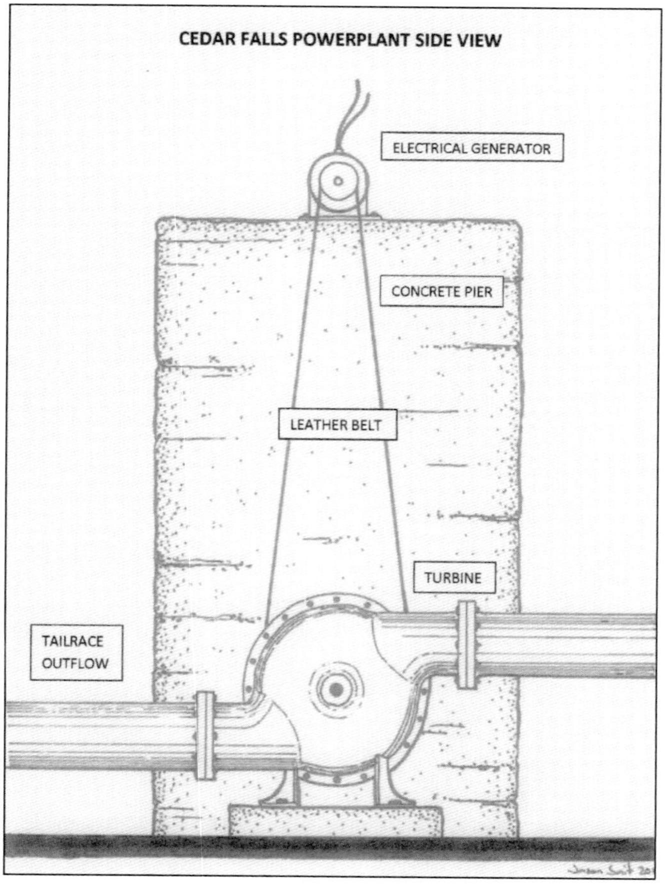

Shubal F. Arnold built a textile mill at Cedar Falls in 1820. It was a small operation, only running 72 spindles. Eventually, it came into the possession of nearby planter Hudson Berry. He and his sons operated it until 1852. Also, there was a general store, water-powered gristmill, sawmill, and cotton gin on the premises. (Courtesy of Greenville County Recreation Department.)

From 1905 to 1940, Cedar Falls Light and Power Company ran an electrical power plant at the falls of the Reedy River. Clarence Williams operated the plant for many years, raising his family across the road. Homes were warned that the lights would soon be turned off when they were flashed on and off. Families would then light their lamps to finish the evening. (Courtesy of Greenville County Recreation Department.)

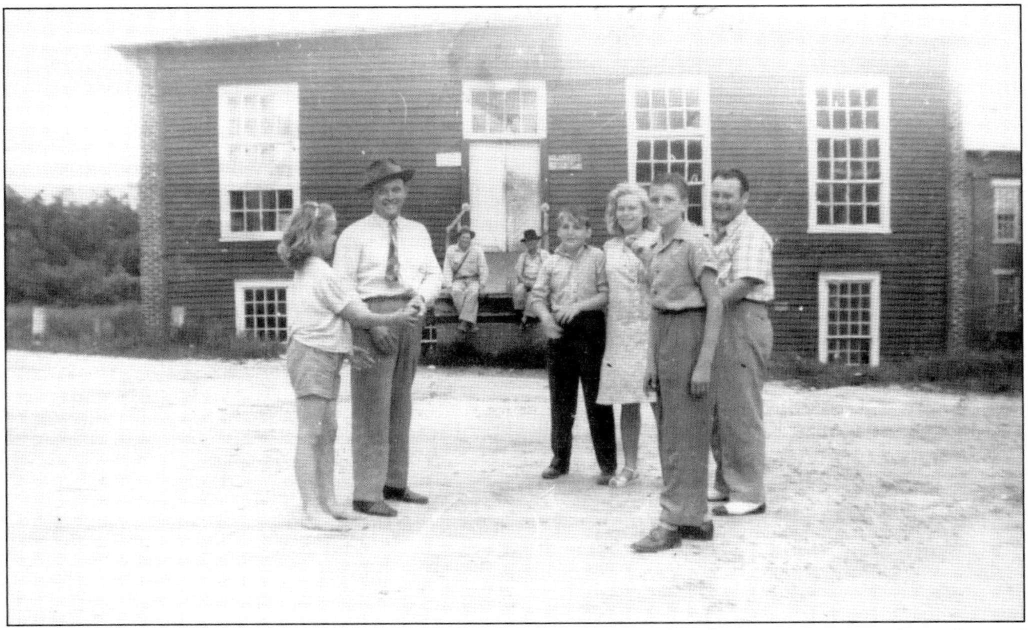

Like most industries, the years during World War II were difficult. This 1942 picture of members of the Williams and Gambrell families reminds people that the mill was the focus of many families' lives. Pictured are, from left to right, Barbara Gambrell, Woodrow Williams, Dennis Williams, Ruth Gambrell, Calvin Williams, and LaVerne Williams. (Courtesy of Doyle Williams.)

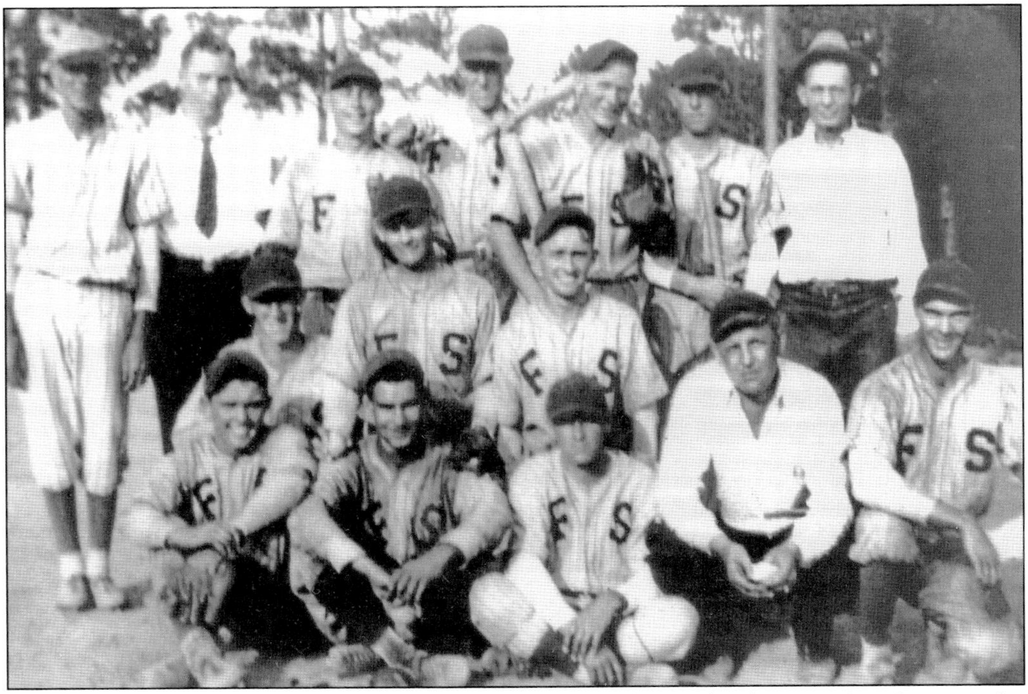

This winning baseball team represented Fork Shoals in the late 1920s. Among the families represented are the Williams, Pitts, Meeks, Smith, Lindley, Thompson, Woods, Gambrell, and Findley. (Courtesy of Dr. Marshall Wiliams.)

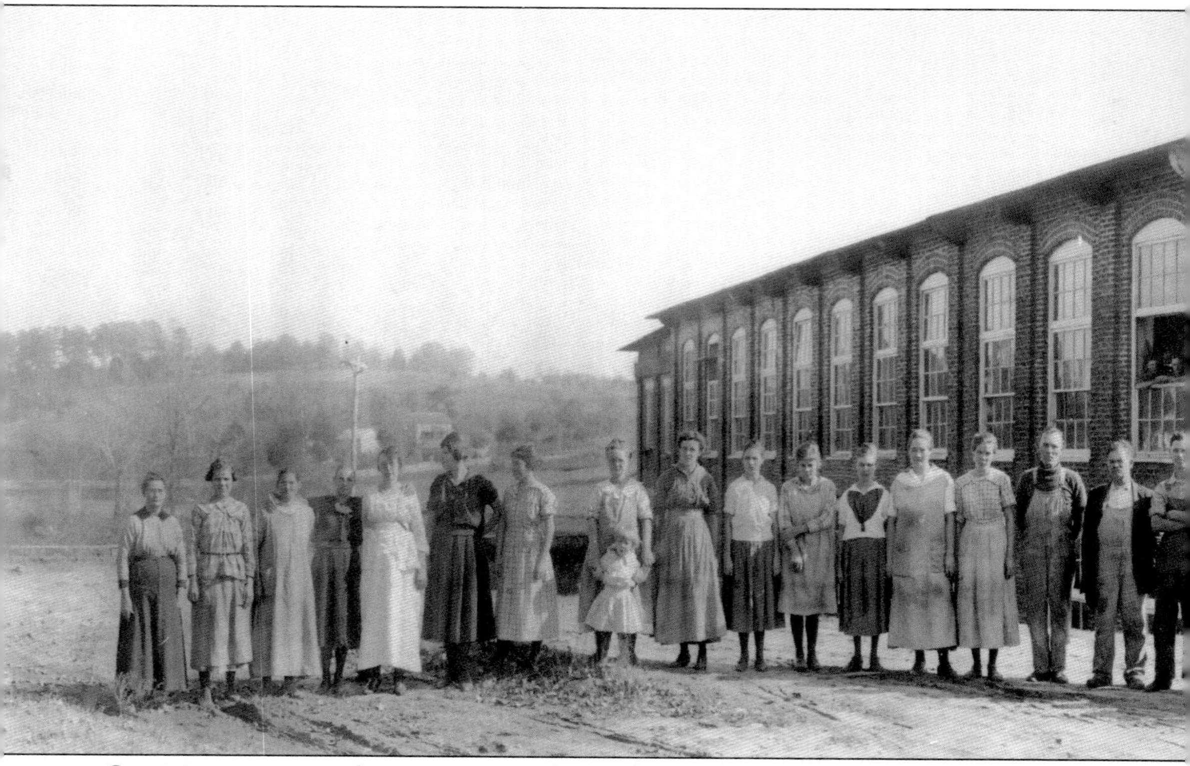

Surviving numerous changes in ownership and natural disasters, the mill grew in the first half of the 20th century. This picture from the mid-1920s shows the growth in employees

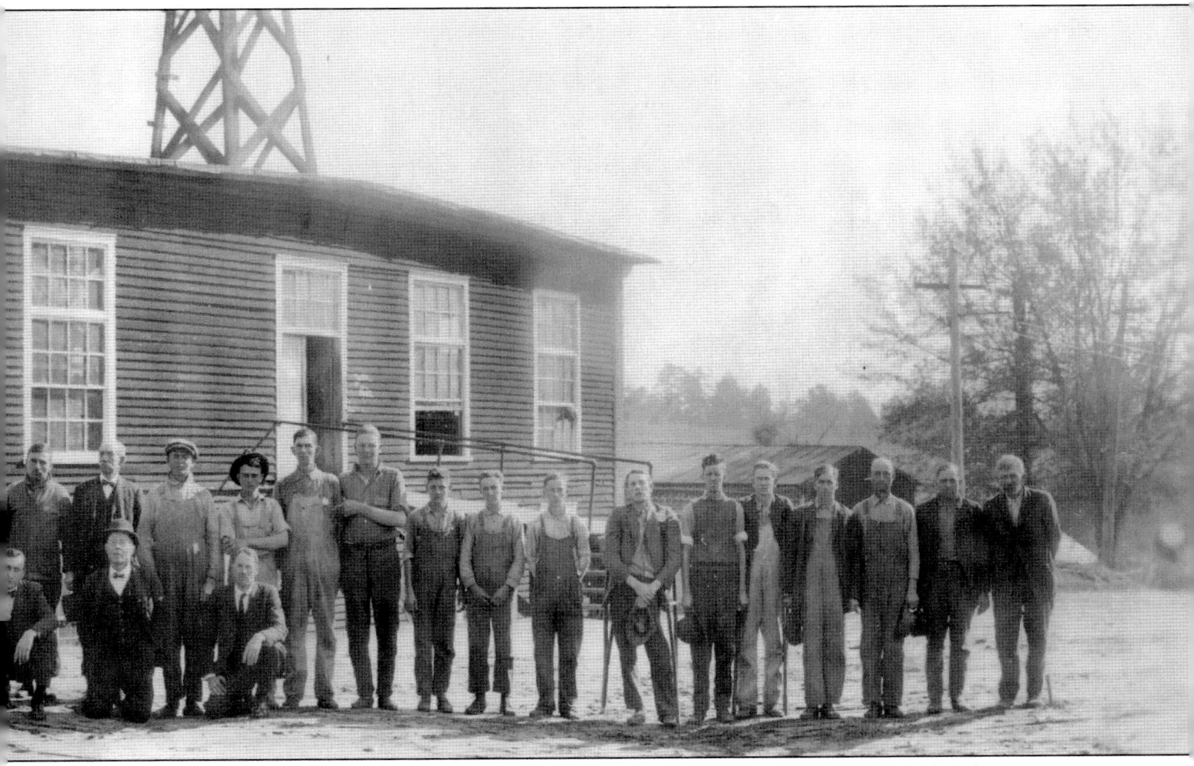

from approximately eight employees 75 years earlier. (Courtesy of Gloria Hughey.)

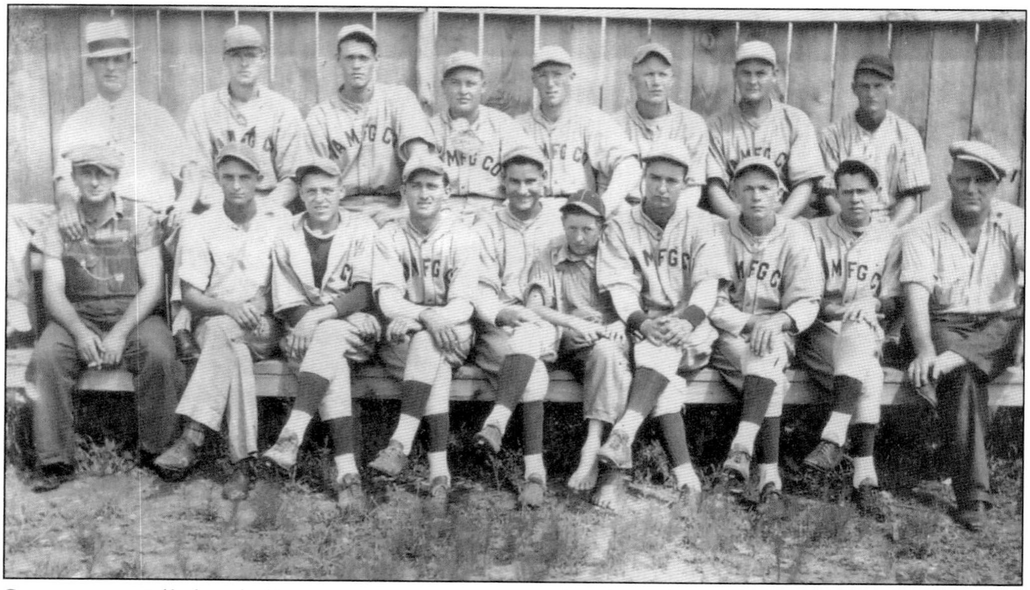

Sports, especially baseball, were an important part of mill life. In the middle decades of the 20th century, Fork Shoals teams were a part of the rich history of textile leagues that covered the Cotton Belt. Familiar faces here include Marvin and Roy Meeks; Harley Moore; JB, Alton, and Leroy Pitts; Calvin Williams; and Johnny Gambrell. (Courtesy of Gloria Hughey.)

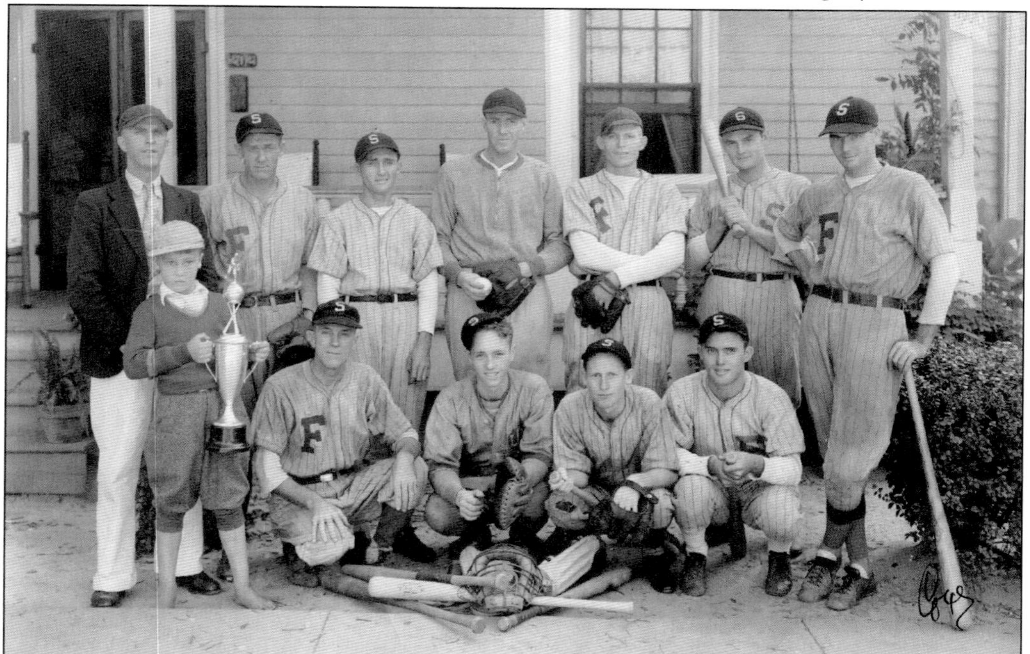

Mill ball was very competitive, and there was tremendous rivalry. There were exposition games with the minor-league teams in the area, and these Textile League teams could hold their own against the professional teams. Many of the Textile League players were drafted by Major League Baseball franchises. The 1933 Fork Shoals Baseball team was the Greenville County League Champions. Some familiar names include Riddle, Thompson, Kirby, Garrison, Williams, Bayne, Woods, Gambrell, and Lindley. (Courtesy of Dr. Marshall Williams.)

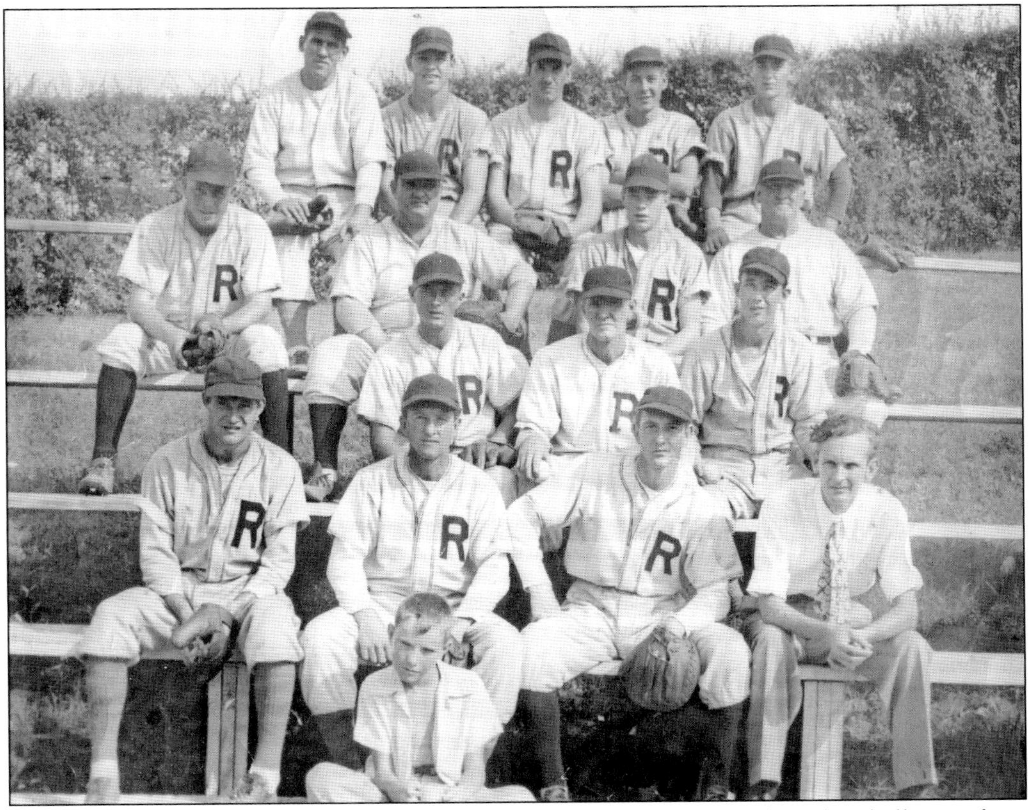

One unique spin-off from this rich sport's history was a community-at-large baseball team from the mid-20th century. Local legend is that this Ridgeway Baseball Team challenged and beat the contemporary Greenville professional team. (Courtesy of Jim Scott.)

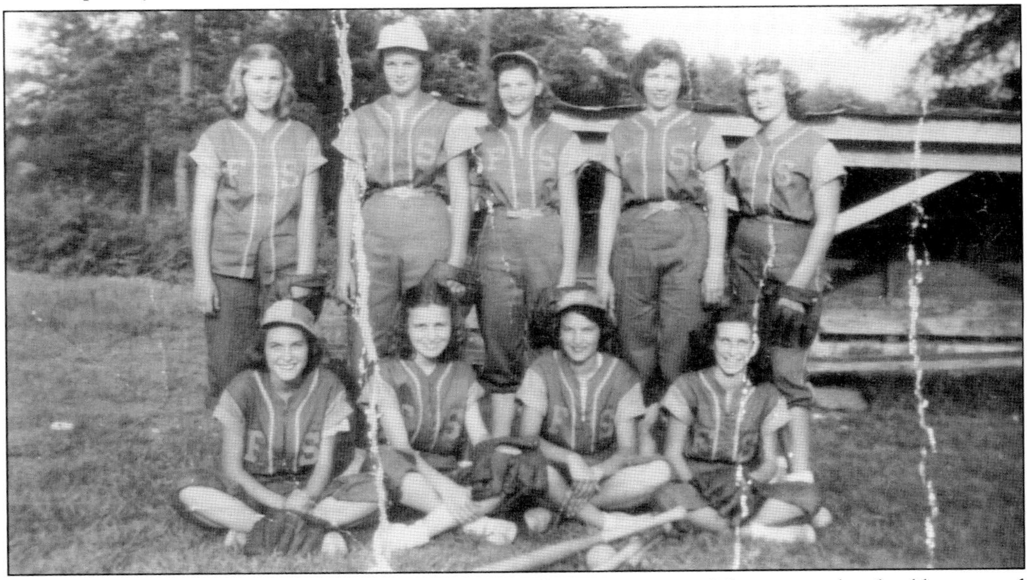

Not to be outdone, the ladies were very present on the sports scene. Mirroring what had happened nationwide during the war years, the ladies had "a league of their own," too. (Courtesy of Gloria Hughey.)

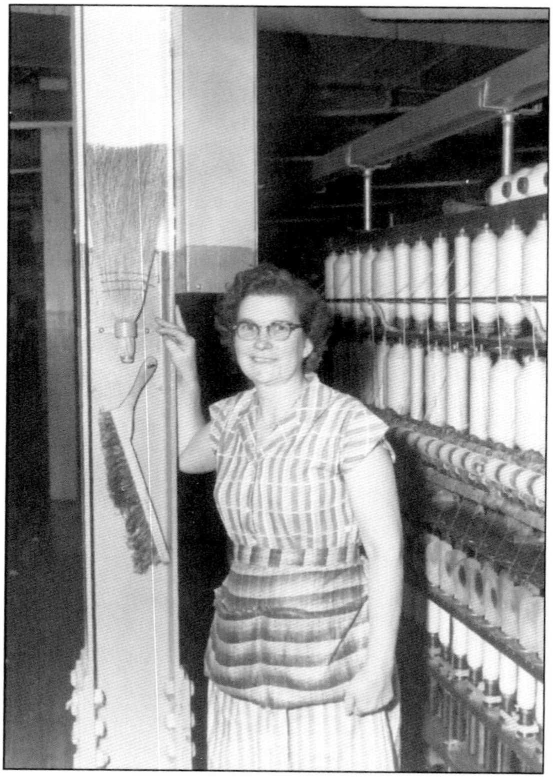

Rose Lee Smith epitomizes the fact that most textile operations were very dependent on ladies. This was definitely true of spinning operations, like Fork Shoals Mill, during almost all of its history. (Courtesy of Janice Smith Woods.)

The standard division of labor was that ladies were responsible for the delicate and tedious spinning operations, while the men were responsible for heavier manual jobs, maintenance, and supervision. From left to right are Willie Ross, Nella Riddle, Mary Ellen Saxon, Leroy Pitts, and Marie King inside Fork Shoals Mill. (Courtesy of Cindy Bradley.)

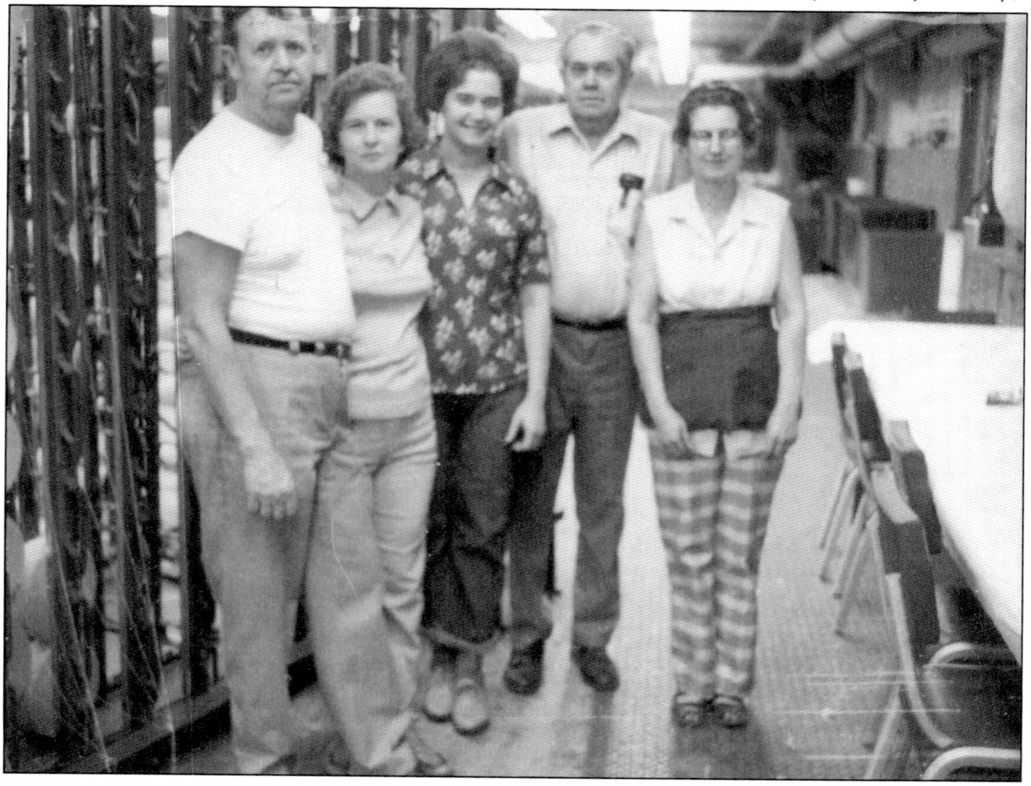

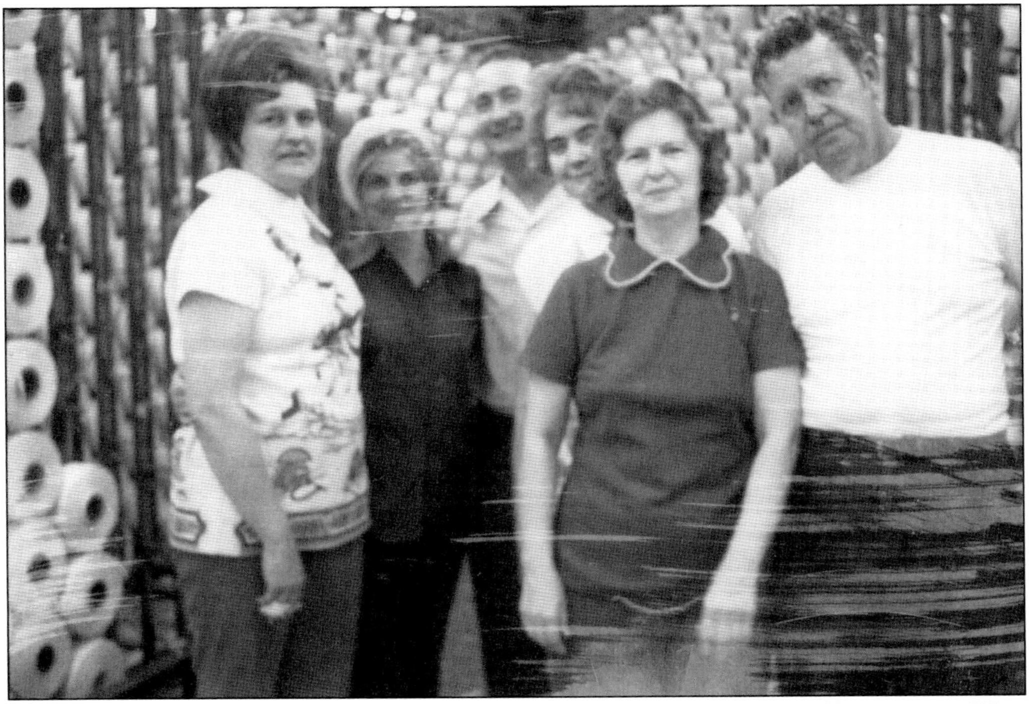

In this snapshot, from left to right, Nancy Kellett, Shirley Bryant, Roy Williams, Mary Ellen Saxon, Nella Riddle, and Willie Ross are standing in front of spools at the mill. (Courtesy of Cindy Bradley.)

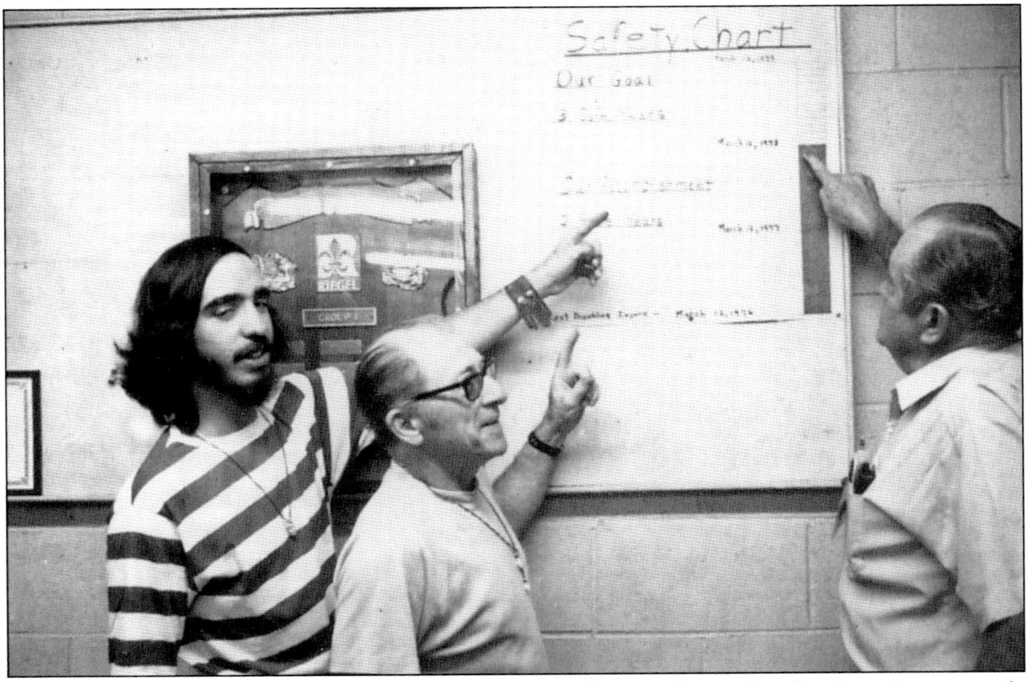

While most mill owners provided safe environments for their employees, legislated regulations made owners, managers, and employees even more aware of the conditions around them. Here, employees are receiving an award for reaching a goal in those efforts. (Courtesy of Gloria Hughey.)

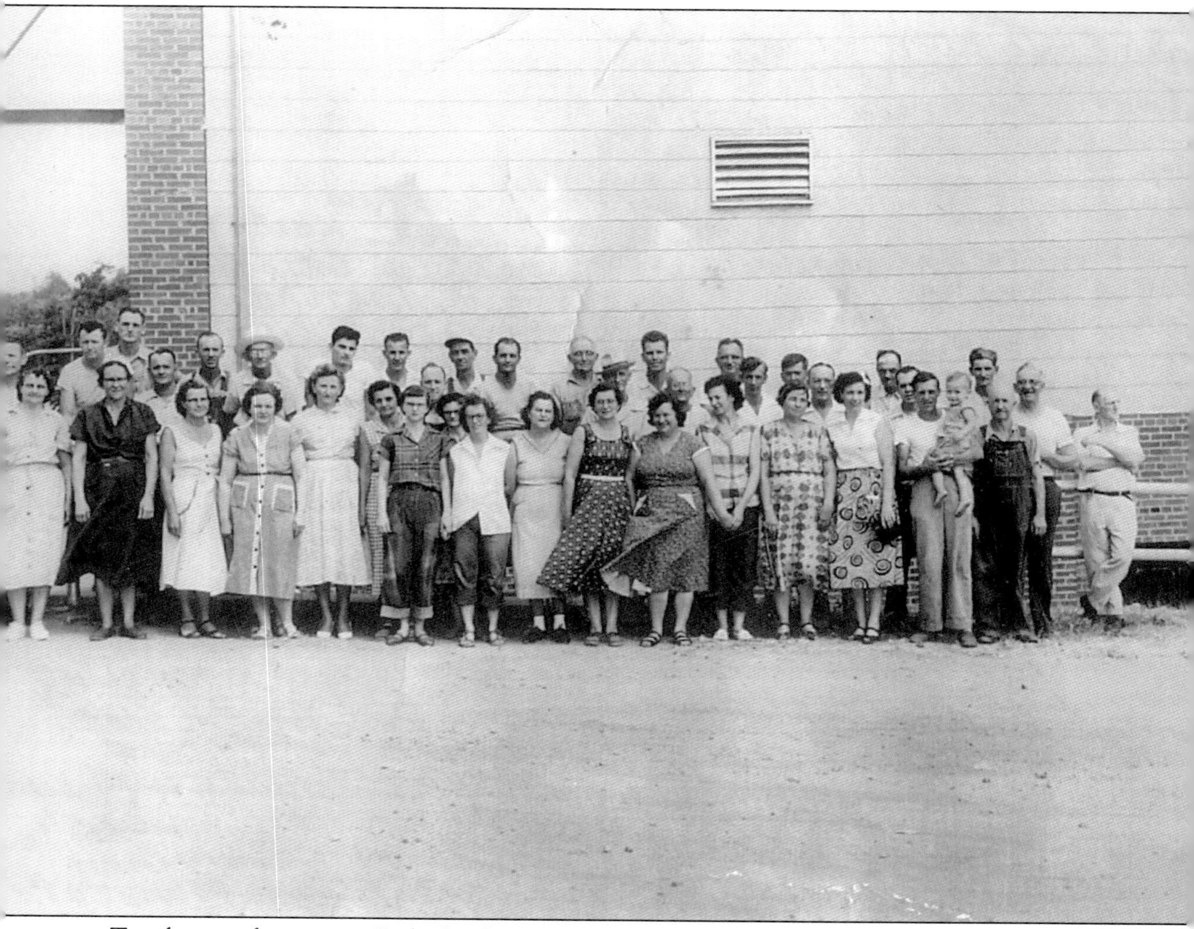

Textile manufacturing at Fork Shoals reached its zenith early in the latter half of the 20th century. This picture shows a relatively large group of employees standing beside a structure that was reasonably modern for its day. As the 20th century wound to a close, textile mills, like the one at Fork Shoals, faced the necessity of competing in a worldwide market. The employees in these pictures from the 1970s were some of the last to battle against the inevitable market forces that eventually closed this historic institution (Courtesy of Janice Smith Woods and Gloria Hughey.)

Five

Growing Roots

Farms

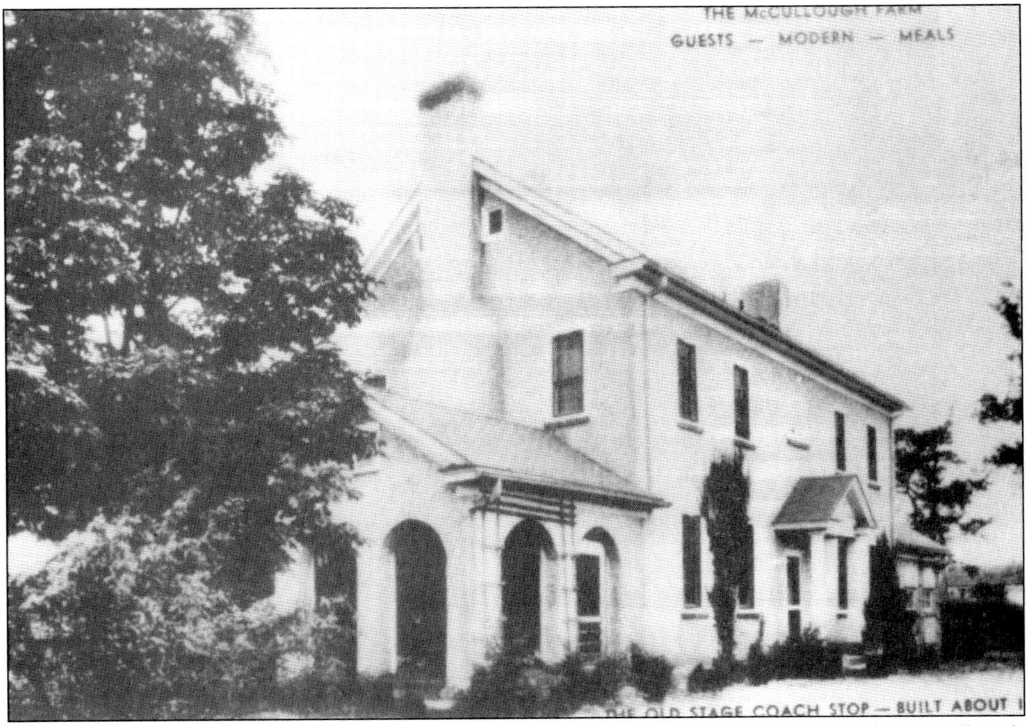

Colonel McCullough was born upon the place where he died along Augusta Road. McCullough was the commander of the famous 16th Regiment, which went into Confederate service on November 19, 1861. A master of Ornan Lodge for many years, McCullough was also engaged in farming and was widely known as progressive and intelligent. (Courtesy of Jim Scott from *A History of the McCullough-Smith Families* by Marie McCullough Derbes.)

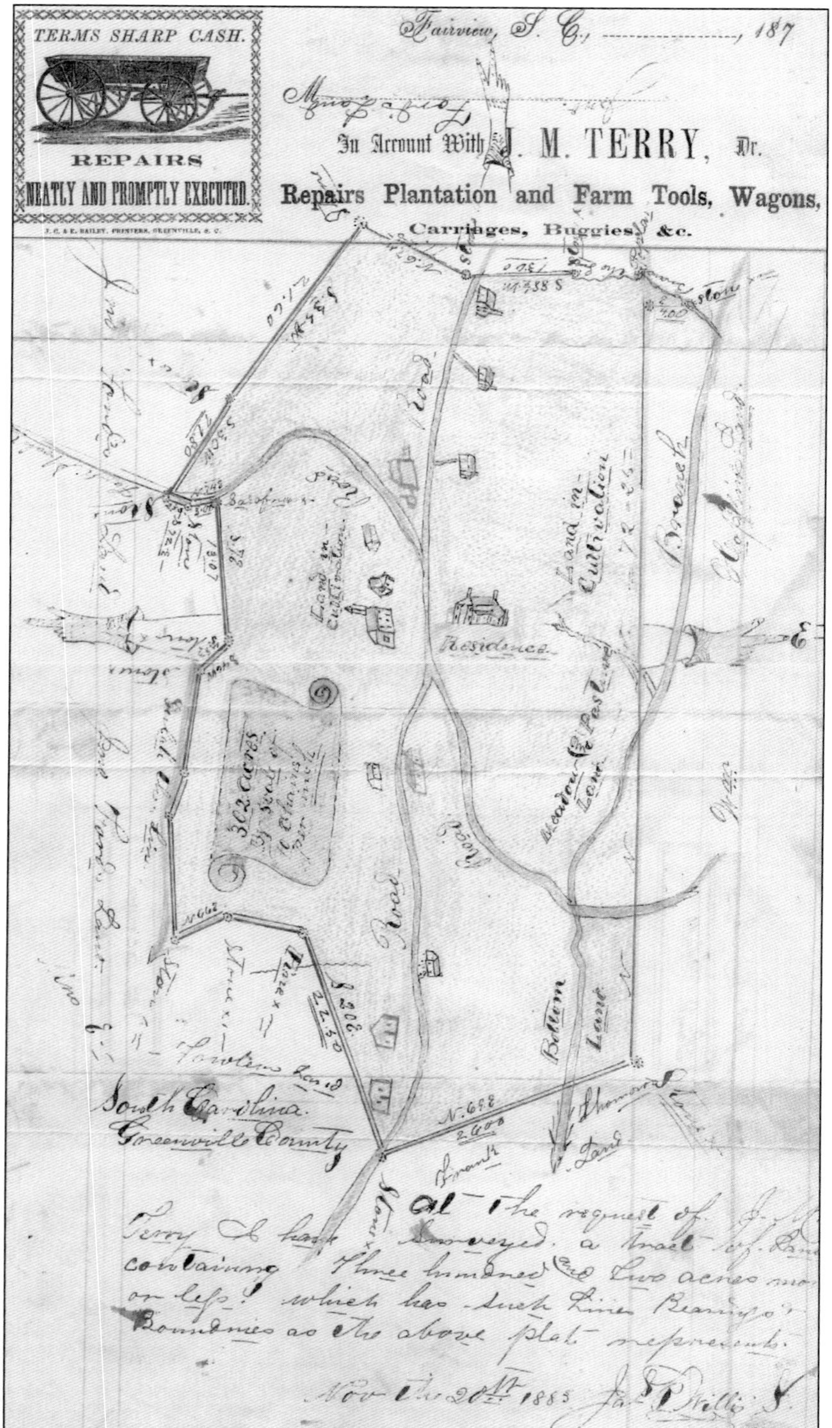

This survey was hand-drawn in 1883 for John Mackin Terry. It artistically shows the 302 acres he owned near the present-day McKelvey and Terry Roads, indicating the location of homes, a church, and other buildings in relationship to bordering owners and natural landmarks. Stones and trees were common property markers, and many still remain today. (Courtesy of Edwin Terry.)

John Mackin "J.M." Terry (1840–1906) and Sallie Hellams Terry (1854–1911), pictured here, were farmers in the Fork Shoals community. John designed the cotton-planter seen below and received a US patent for it on January 12, 1892. "I have invented certain new and useful Improvements in Cotton-Planters; and I do declare the following to be a full, clear, and exact description of the invention, such as will enable others skilled in the art to which it appertains to make and use the same." (Right, courtesy of Edwin Terry; below, courtesy of the US Patent Office.)

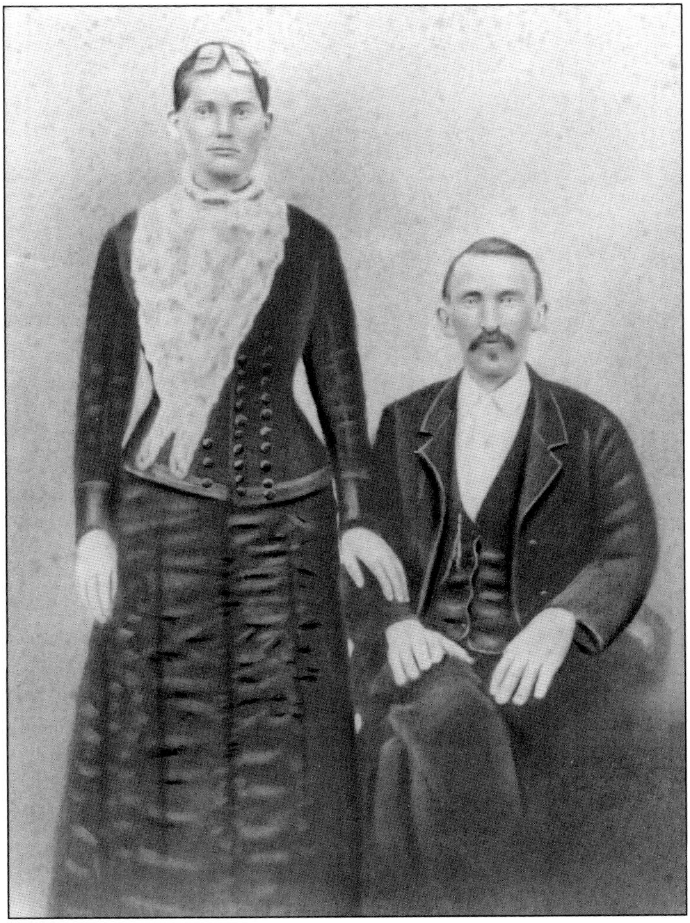

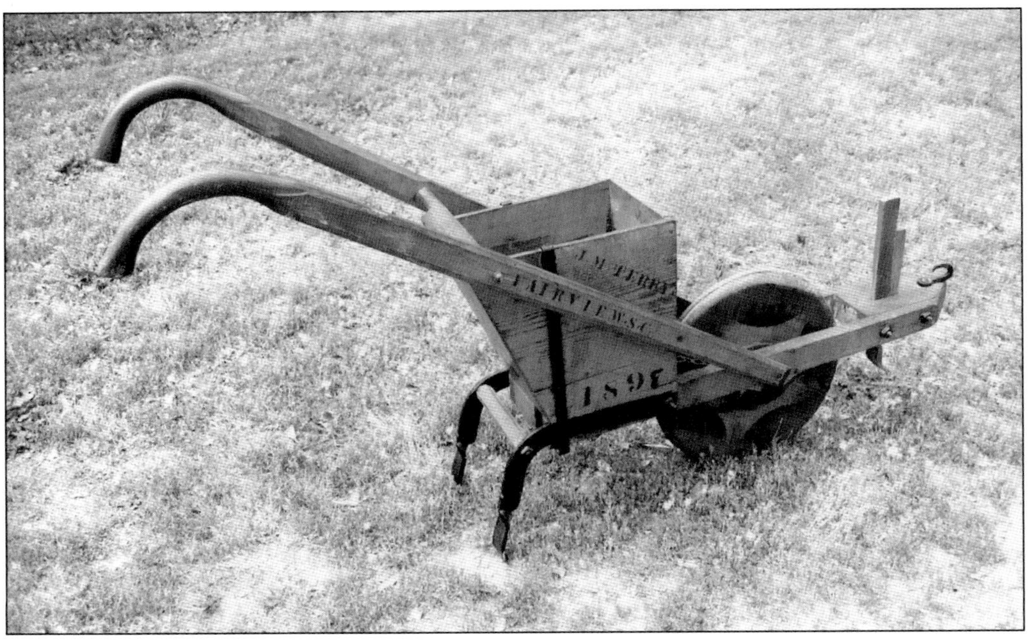

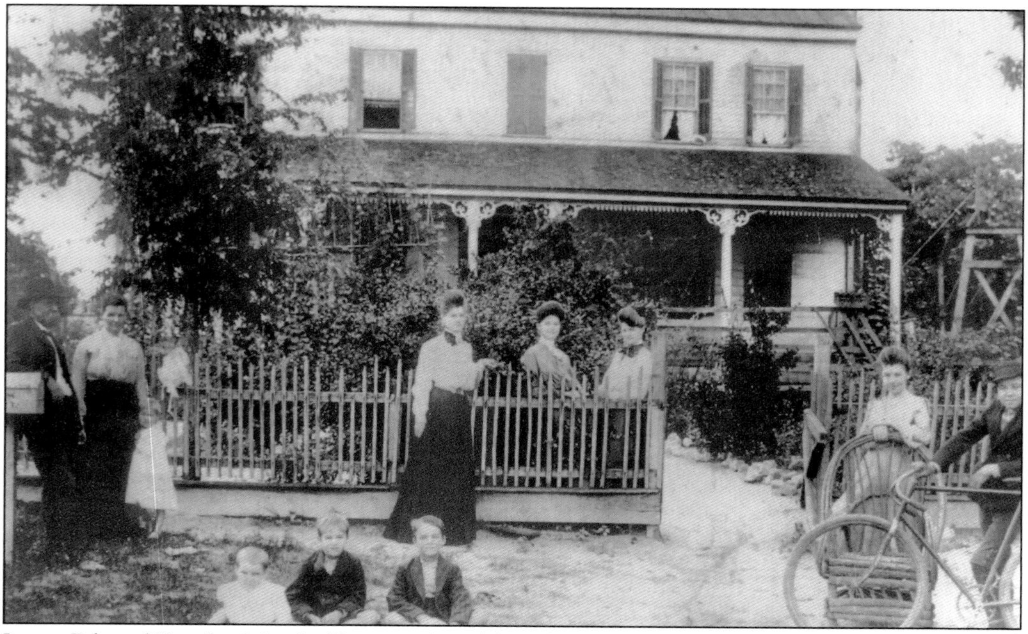

James Edward Knight, Martha Eugenia Arnold, and family pose before the home they purchased (1880s) for 10 bales of cotton weighing 500 pounds each and worth 5¢ per pound. Knight was a sawyer often buying land for the timber. Ed Knight cut the timber and helped build many structures and homes in the area, including the original part of Sandy Springs Baptist Church. (Courtesy of John Drayton Hopkins's Family.)

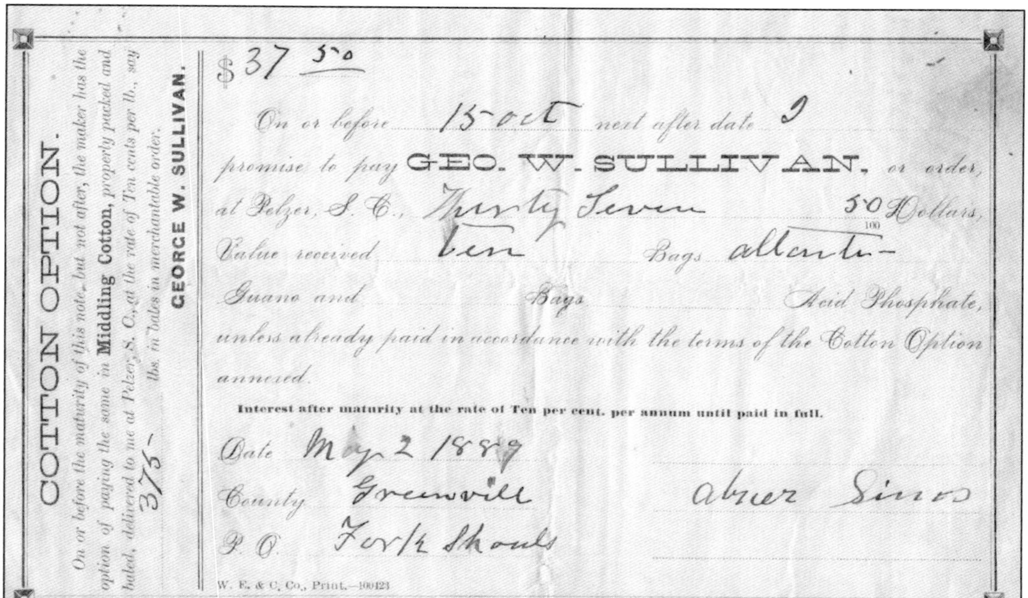

COTTON OPTION.

On or before the maturity of this note, but not after, the maker has the option of paying the same in **Middling Cotton**, properly packed and baled, delivered to me at Pelzer, S. C., at the rate of Ten cents per lb., say 375 lbs. in bales in merchantable order.

GEORGE W. SULLIVAN.

$37 50/100

On or before 15 oct next after date I promise to pay GEO. W. SULLIVAN, or order, at Pelzer, S. C., Thirty Seven 50/100 Dollars, Value received Ten Bags Atlanta Guano and Bags Acid Phosphate, unless already paid in accordance with the terms of the Cotton Option annexed.

Interest after maturity at the rate of Ten per cent. per annum until paid in full.

Date May 2 1889

County Greenvill

P. O. Fork Shoals

Abner Sims

W. F. & C. Co., Print.—10423

Cotton was a commodity. Fertilizer was a necessity. Fertilizer could be obtained from the local cotton mill, and farmers could opt to pay this debt with cotton bales at harvest. George Washington Sullivan of Sullivan Manufacturing Company managed this transaction, dated 1889. Cotton was graded according to quality and color. The grading scale term "middling" could be middling fair, good middling, or low middling. (Courtesy of Cheryl MacKnight.)

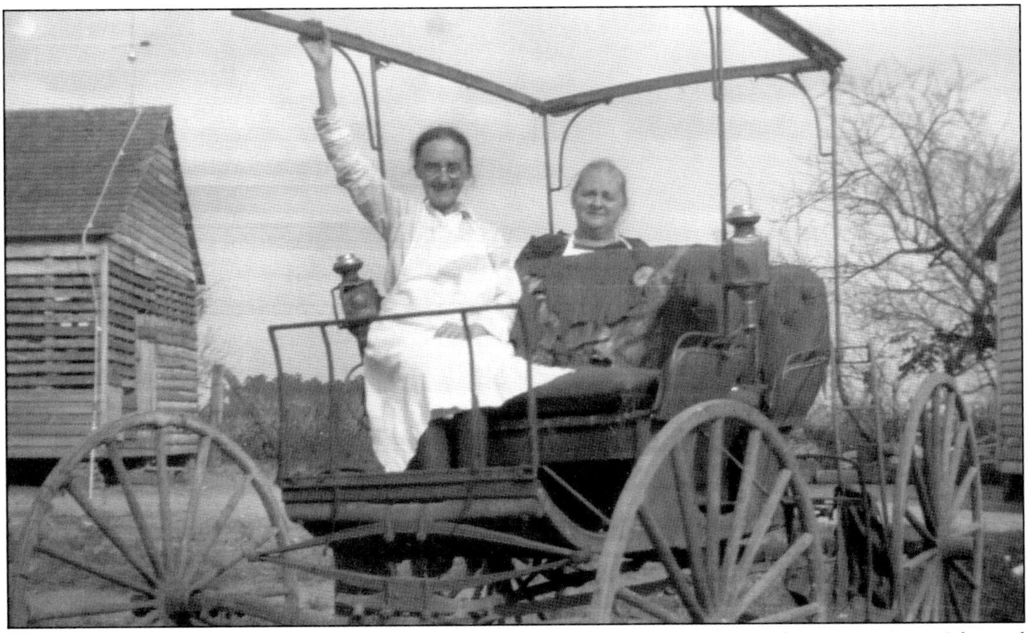

Horse-drawn carriages were common modes of transportation in the early 1900s. Aunt Ida and Granny Beulah Vaughan Riddle (great-great-grandmother of resident Clyde Jenkins) look like they have been working hard and appreciate this respite with their buggy. Walking everywhere was a typical means of getting around. It has been said that a visit nine miles away was nothing; one could get there and back in a day. (Courtesy of Clyde Jenkins.)

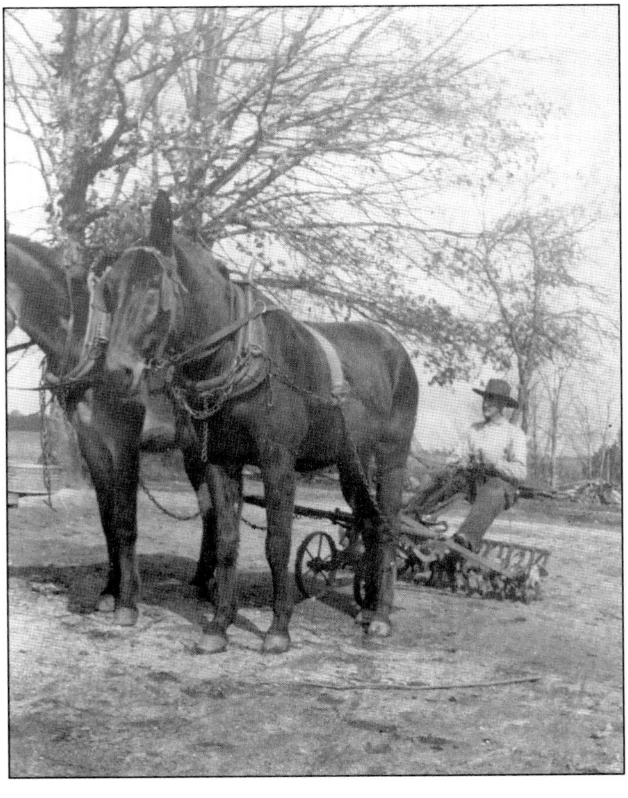

Mules were popular and dependable additions to the farms. Clyde Jenkins's great-grandfather Jimmy Riddle is preparing to plow with his hardworking beasts at the Riddle home place on Fork Shoals Road. In days past, mules were sometimes used to operate grain mills, or grist stones, before the water mills were operating along the Reedy River. (Courtesy of Clyde Jenkins.)

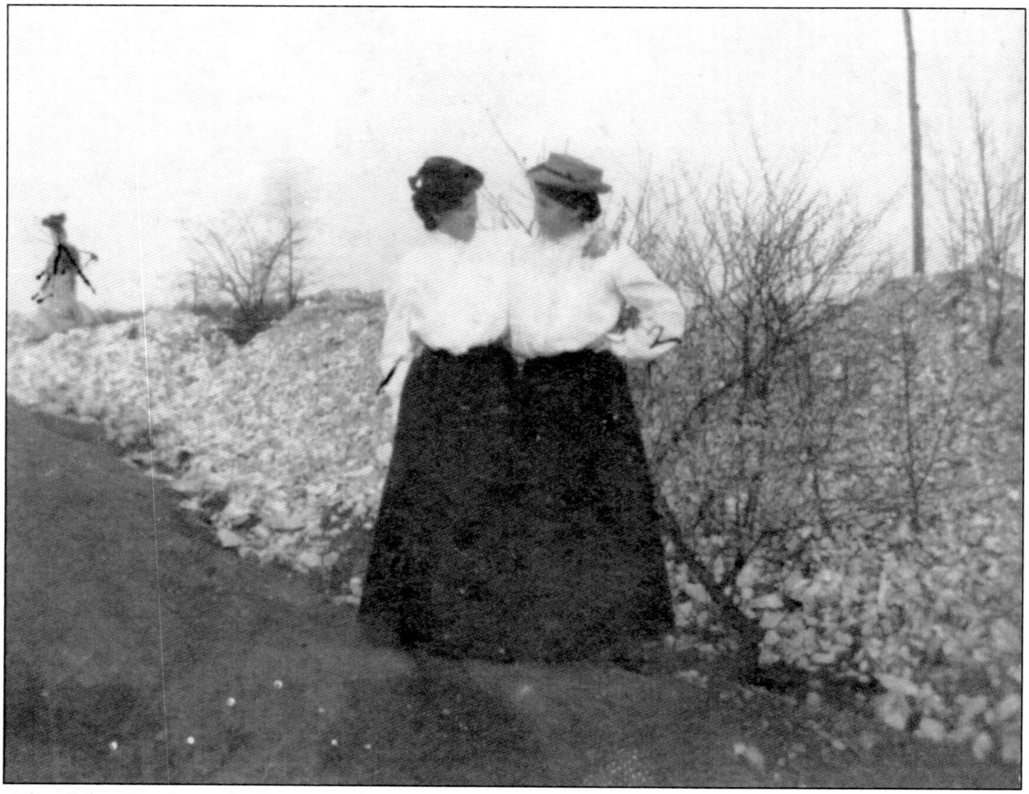

After John Hopkins died in 1837, his widow, Lucinda Hopkins (1800–1876), managed their farm. Thus began a long period of farm ownership and control by the women in the Hopkins family. This photograph shows Nell Knight Hopkins (left) and her sister Jean Knight Gilliland posing in front of their cotton fields. (Courtesy of John Drayton Hopkins's Family.)

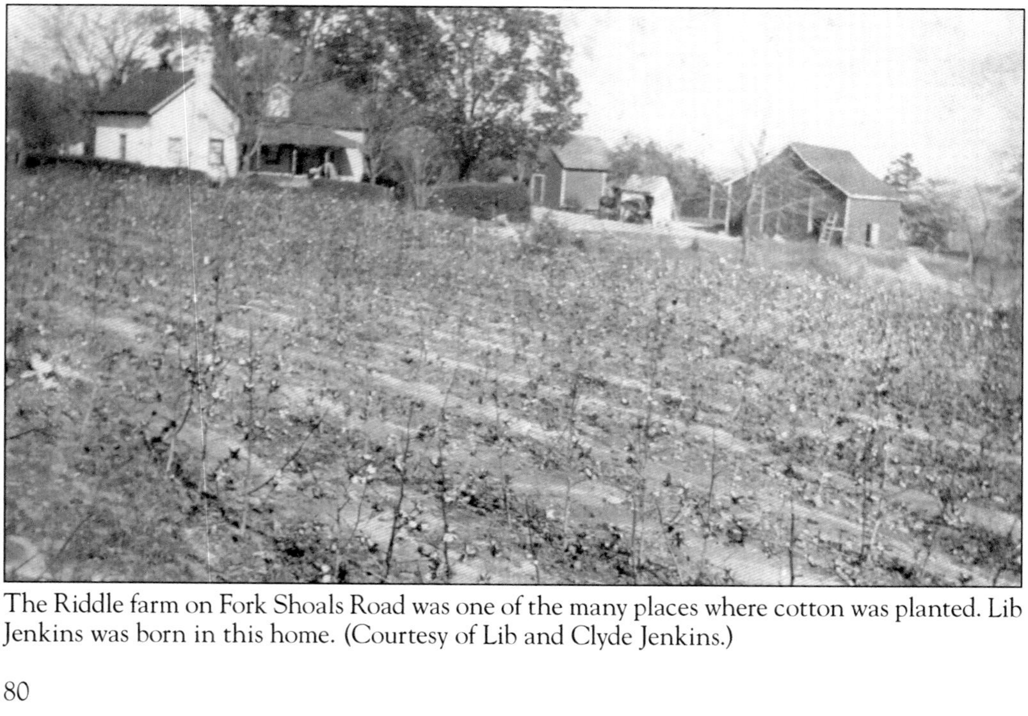

The Riddle farm on Fork Shoals Road was one of the many places where cotton was planted. Lib Jenkins was born in this home. (Courtesy of Lib and Clyde Jenkins.)

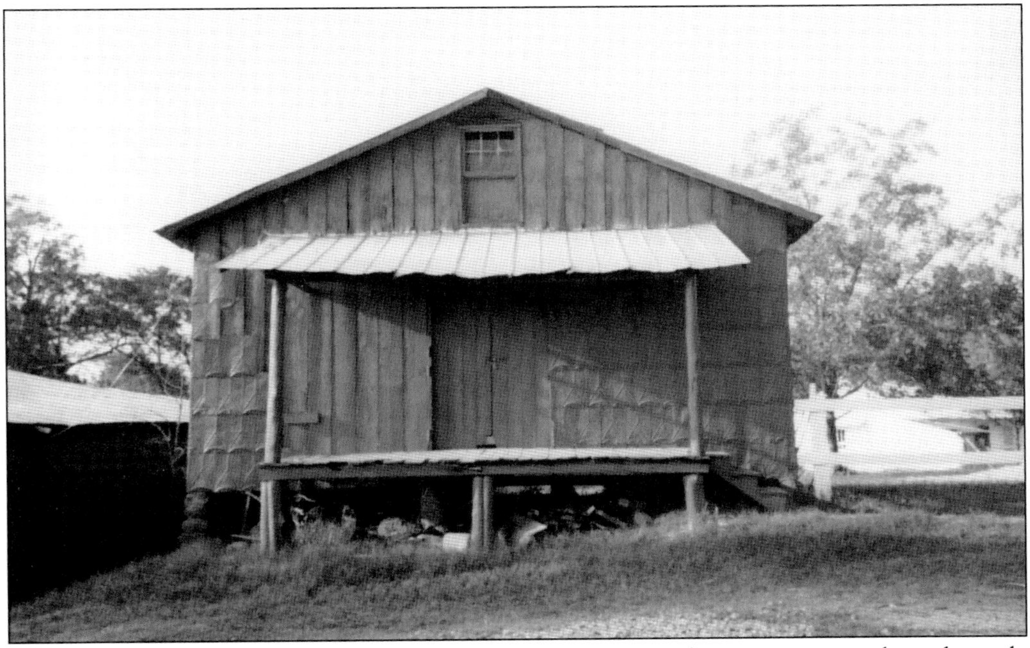

This stamped tin–clad barn was originally used to store cotton that was waiting sale in the early to mid-1900s. Even as late as 1987, there were remnants scattered throughout the several bays inside. Many properties in this area, old family farms and old family estates, still contain artifacts of bygone days. (Courtesy of Carol Gilley.)

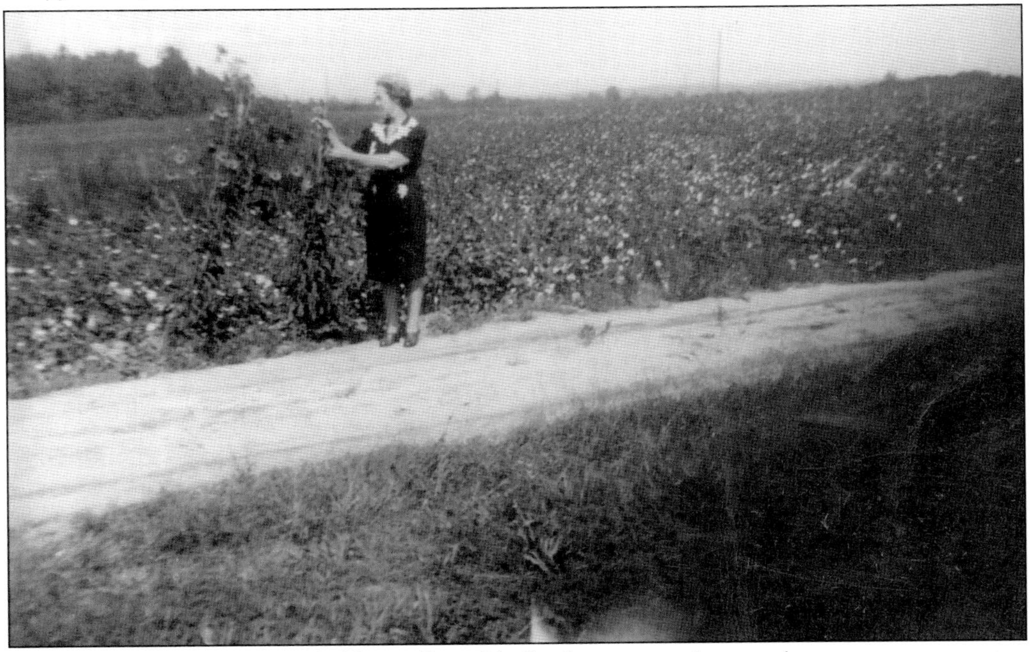

Lula Abercrombie Tumblin held a stalk and boll of cotton as she posed next to a vast cotton field along Fork Shoals Road in the 1940s. Improved methods of cultivation were propagated by agricultural societies at this time, the first one established in Philadelphia. The second one in the nation was in South Carolina, and it still operates under its original charter. (Courtesy of Lou Ann Baldwin.)

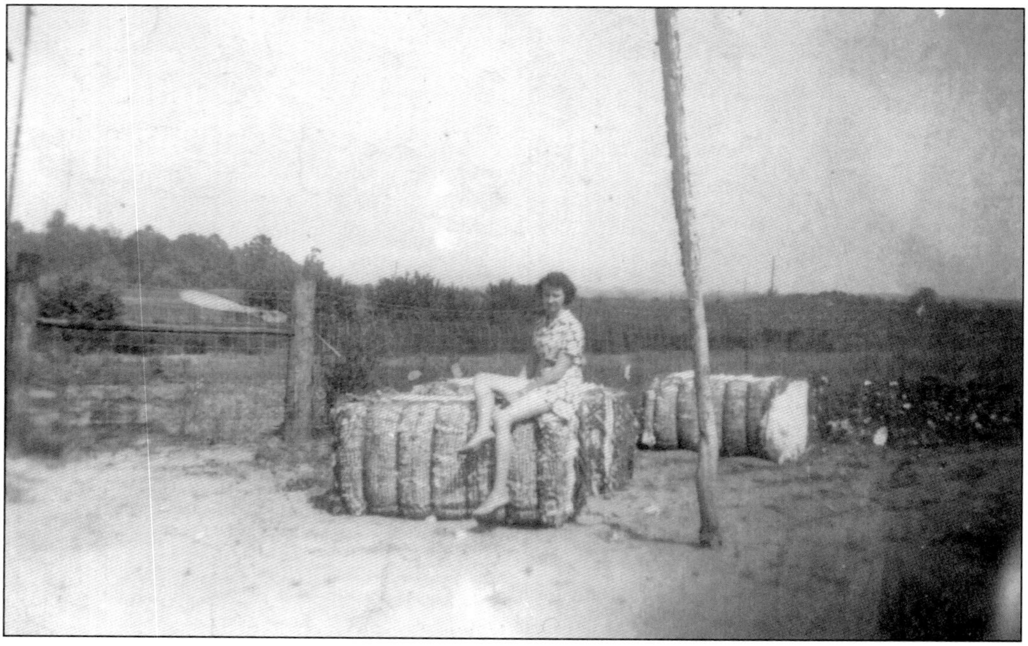

Sara Tumblin Holt sat atop a bale of cotton near Oaklawn and Fork Shoals Roads. In the 1940s, cotton was still widely planted by those with large farming operations as well as by those with smaller plots. Cotton was no longer king, but it was still a profitable crop that could be sold. (Courtesy of Lou Ann Baldwin.)

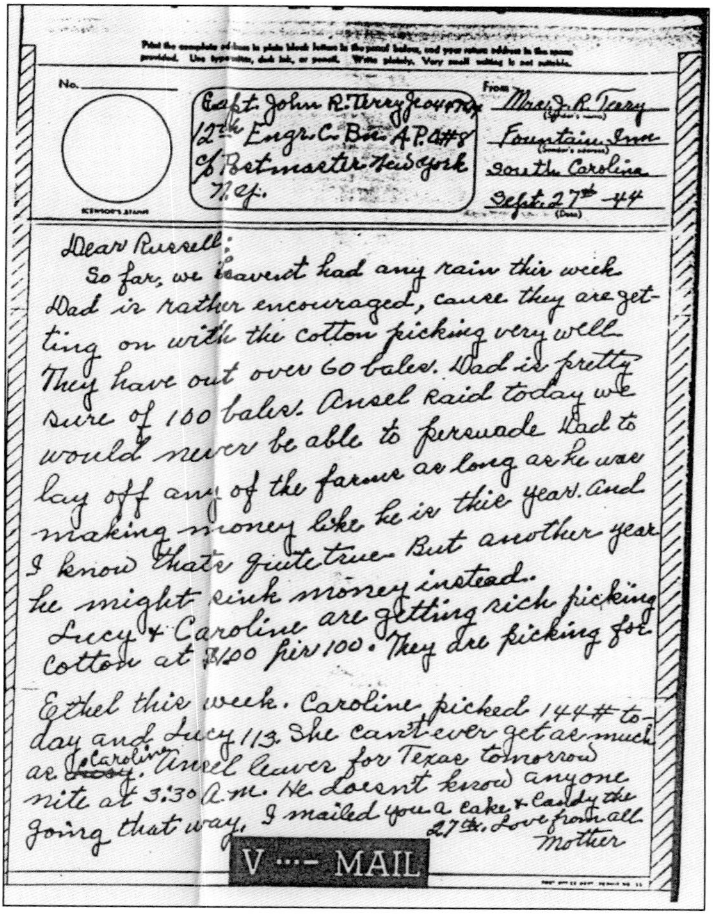

No.

Capt. John R. Terry Jr.
12th Engr. C. Bn. A.P.O. #8
c/o Postmaster New York
N.Y.

From Mrs. J. R. Terry
Fountain Inn
South Carolina
Sept. 27th 44

Dear Russell:
So far, we haven't had any rain this week. Dad is rather encouraged, cause they are getting on with the cotton picking very well. They have out over 60 bales. Dad is pretty sure of 100 bales. Ansel said today we would never be able to persuade Dad to lay off any of the farms as long as he was making money like he is this year. And I know that's quite true. But another year he might sink money instead.
Lucy & Caroline are getting rich picking cotton at $1.00 per 100. They are picking for Ethel this week. Caroline picked 144# today and Lucy 113. She can't ever get as much as ~~Lucy~~ Caroline. Ansel leaves for Texas tomorrow nite at 3:30 A.M. He doesn't know anyone going that way. I mailed you a cake & candy the 27th. Love from all.
Mother

V ···– MAIL

The Terry family saved this "victory letter" from World War II. Written to her soldier, this mother writes of everyday farming in Fork Shoals, including how many pounds Ethel and Caroline picked in a day. (Courtesy of the Terry family.)

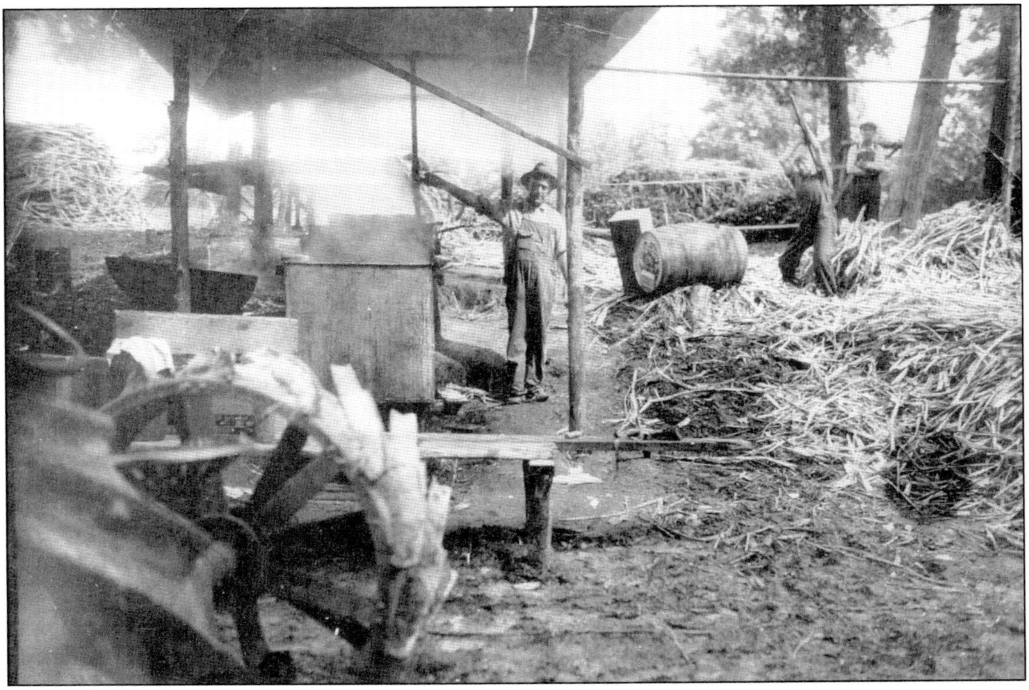

Jim Terry ran this molasses mill near the creek on his farm on McKelvey Road. Stalks of sorghum cane, commonly known as "'lasses cane," were processed here to produce a popular and delicious sweetener. This photograph was taken in the 1930s. Locally grown sugar was a desired product during the Depression years. (Courtesy of Shirley Brashier Stone)

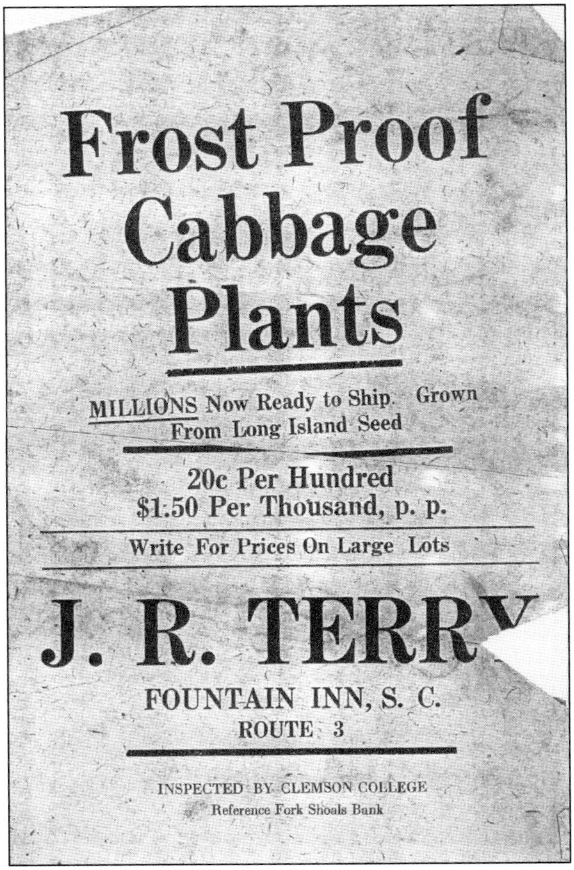

Frost Proof
Cabbage
Plants

MILLIONS Now Ready to Ship. Grown
From Long Island Seed

20c Per Hundred
$1.50 Per Thousand, p. p.

Write For Prices On Large Lots

J. R. TERRY

FOUNTAIN INN, S. C.
ROUTE 3

INSPECTED BY CLEMSON COLLEGE
Reference Fork Shoals Bank

This flier advertised frost-proof cabbage plants grown and sold by J.R. Terry in the 1930s and 1940s. His crop was bought locally as well as shipped throughout the United States. Many advances were made, and farms prospered. This was the beginning of the agricultural boom that lasted until the early 1970s. (Courtesy of Edwin Terry.)

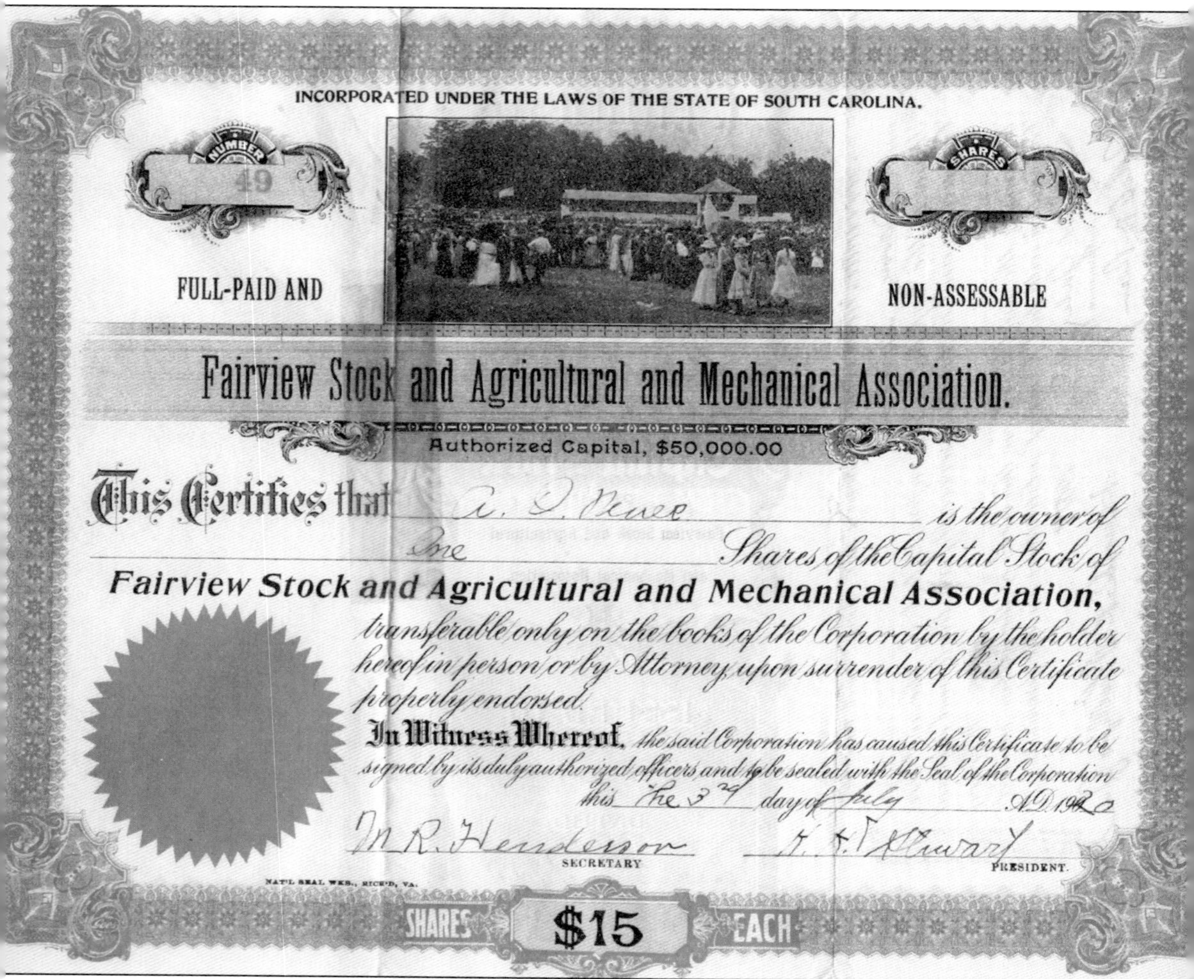
INCORPORATED UNDER THE LAWS OF THE STATE OF SOUTH CAROLINA.

NUMBER 49

SHARES

FULL-PAID AND

NON-ASSESSABLE

Fairview Stock and Agricultural and Mechanical Association.

Authorized Capital, $50,000.00

This Certifies that [illegible] is the owner of One Shares of the Capital Stock of

Fairview Stock and Agricultural and Mechanical Association,

transferable only on the books of the Corporation by the holder hereof in person or by Attorney upon surrender of this Certificate properly endorsed.

In Witness Whereof, the said Corporation has caused this Certificate to be signed by its duly authorized officers and to be sealed with the Seal of the Corporation this The 3rd day of July A.D. 1920

M. R. Henderson
SECRETARY

[illegible]
PRESIDENT.

NAT'L SEAL WKS., RICH'D, VA.

SHARES $15 EACH

Shares of stock were sold for part ownership of the Fairview Stock Association. Only men could buy these shares. This one was purchased in 1920 and handed down to a male heir. Breeders were becoming more organized into the 1900s. The buying and selling of livestock was managed and was of vital interest to the cattlemen and livestock producers. (Courtesy of Carolyn Martin Shirley.)

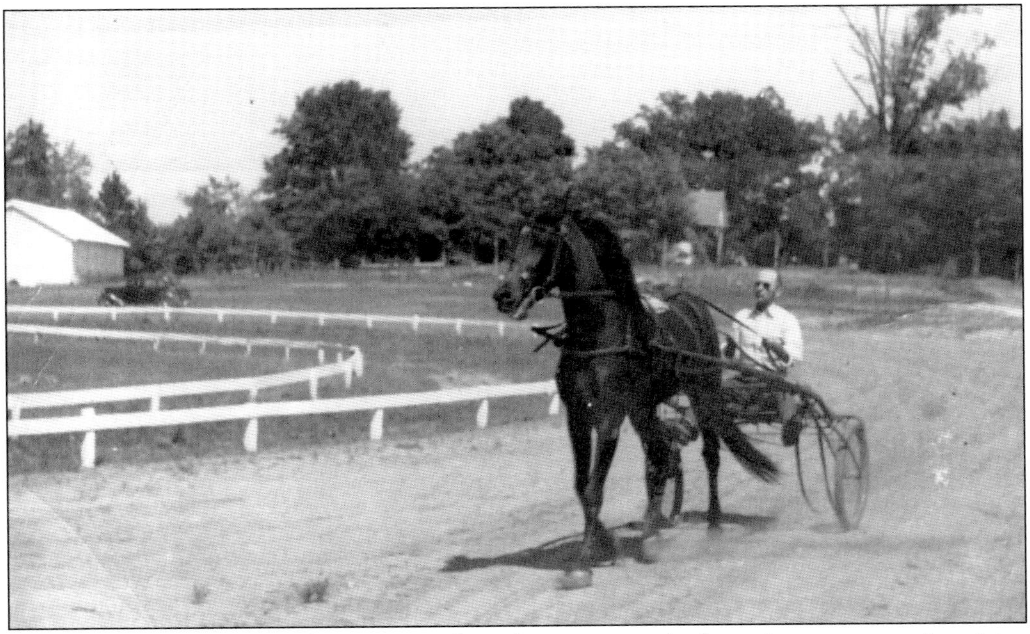

The fairgrounds were a vital connection for information on the latest improvements in raising crops and animal husbandry. Information on marketing techniques and agricultural advancements were shared. It was an important community event, mixed with a lot of fun and competitions. Harness racing was a popular sport, and the Fairview Stock Show Fairgrounds had a nice track. Here, Murphy Phillips is shown practicing in the 1930s. (Courtesy of Edwin Terry.)

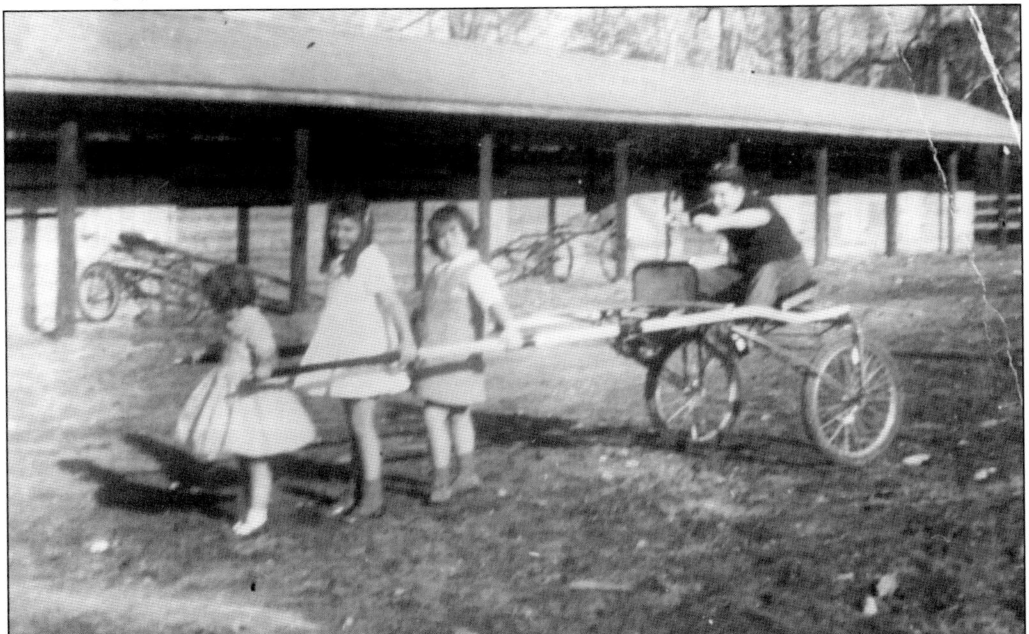

These children enjoyed pretending to harness race at the Fairview Stock Show Fairgrounds in the 1960s. The driver's name was David Phillips, and the girls were (from left tor right) Anna Marie Martin Bryant, Anna Marie Nash, and Pam Phillips, as "the horses." The fair brought folks together for a break from the regular work routine and school. (Courtesy of Pam Phillips.)

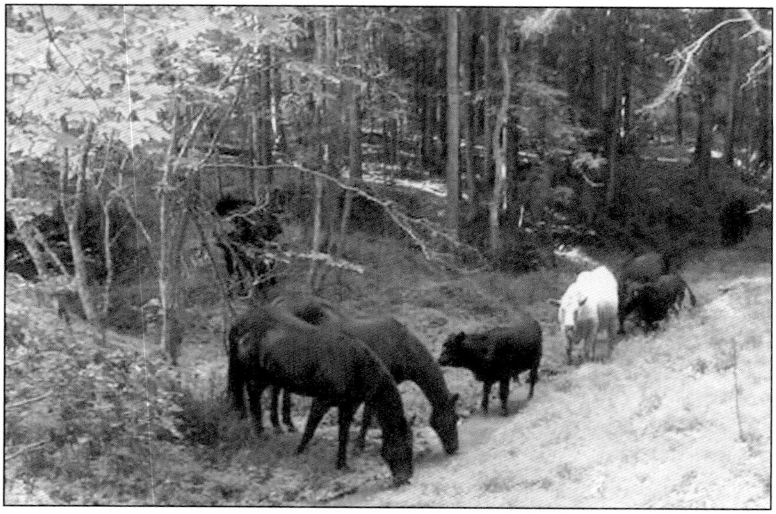

Horses and cattle have been a part of the landscape at Fork Shoals for hundreds of years. It is still a rural area, and many sights such as this are common. Also, much farming is still practiced with farmers cutting and baling hay for these animals. (Courtesy of Carol Gilley.)

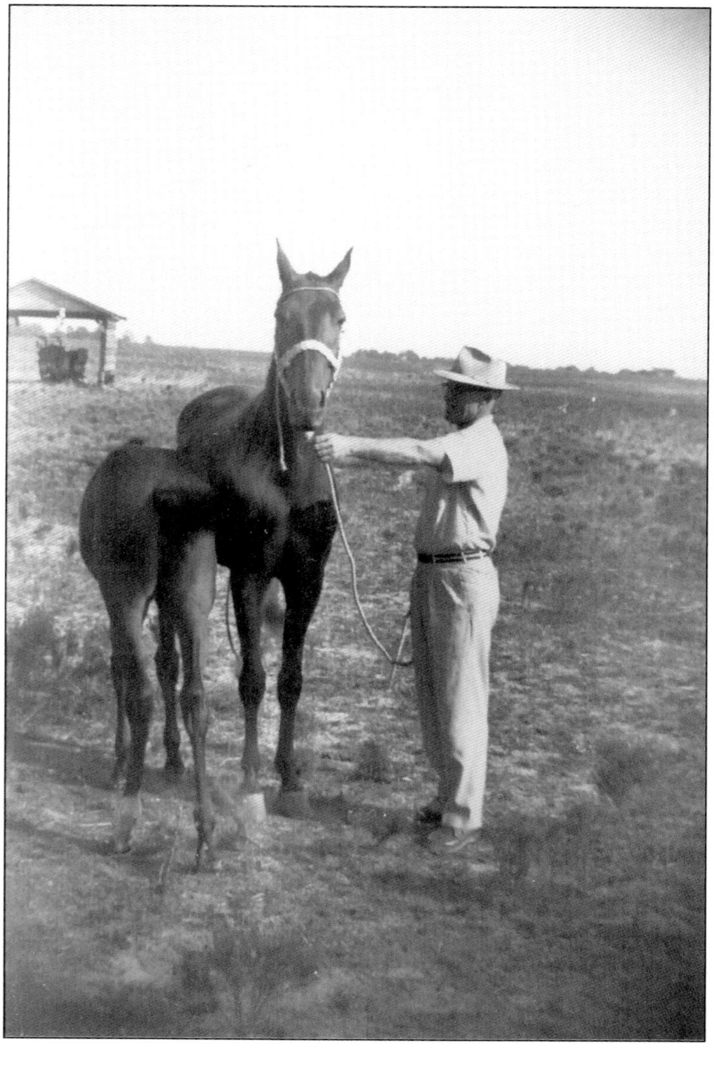

On his farmland alongside McKelvey Road, Albert O'Hair Neves proudly showed off his racehorse "Miss O'Hair" as she nursed her foal in 1954. He enjoyed harness racing with her at the Fairview Stock Show Fairgrounds. The mare got a break from training for a while. The foal would likely become another prized working companion for its owner. (Courtesy of Margaret Seel.)

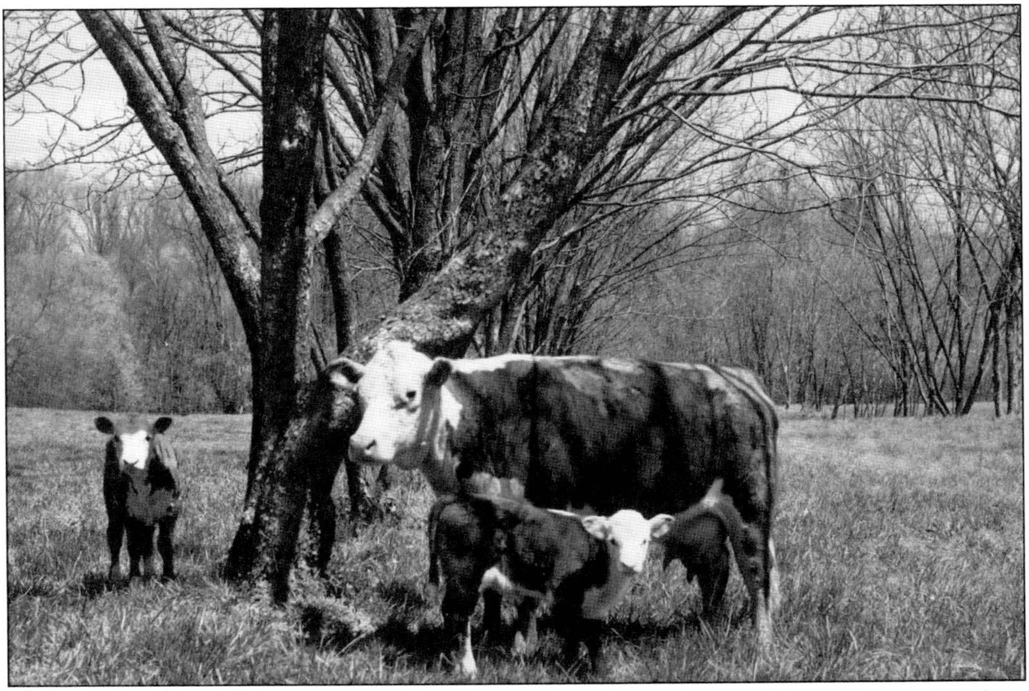

These Hereford cattle are grazing in Patriot's Grove along the banks of the Reedy River on Hopkins property. Patriot's Grove is a grove of pecan trees that were planted in the 1800s by the Hopkins to remember the soldiers of the Civil War. (Courtesy of John Drayton Hopkins's Family.)

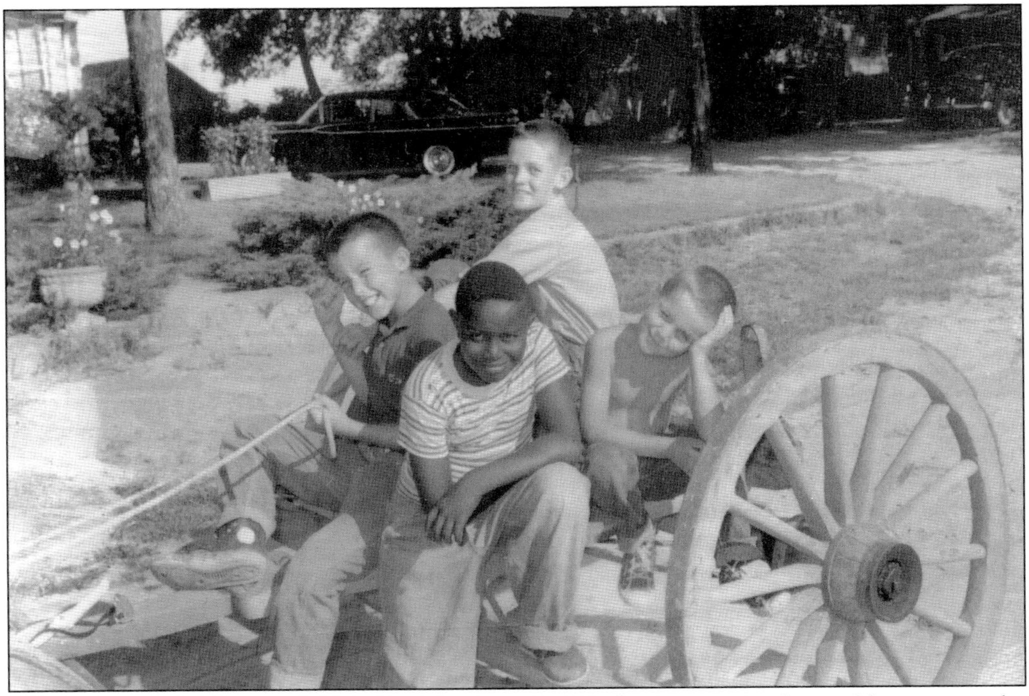

These friends enjoy playing on the Robert Arnold Knight property with an old wagon in the 1950s. Billie Gaines is in the foreground and in the back are, from left to right, Tommy Baldwin, John Agee, and Eddie Agee. (Courtesy of Virginia Agee Dean.)

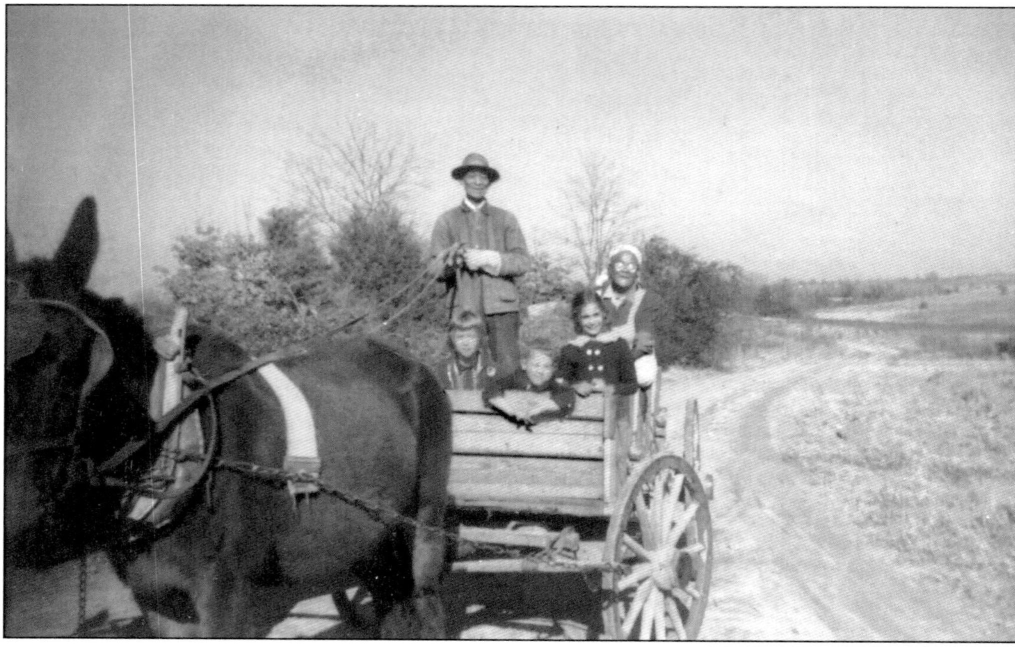

There is still a large number of old family estates in southern Greenville County. A tenant farmer and the household cook, known as Uncle Will and Aunt Neal, took Leonard Allbritton (front, middle), Patricia Allbritton (front, left), and cousin Virginia Agee Dean (front, right) on a regular excursion by wagon. In the 1950s, a favorite way of traveling between the old cotton gin and the home was still by horse and wagon. (Courtesy of Virginia Agee Dean.)

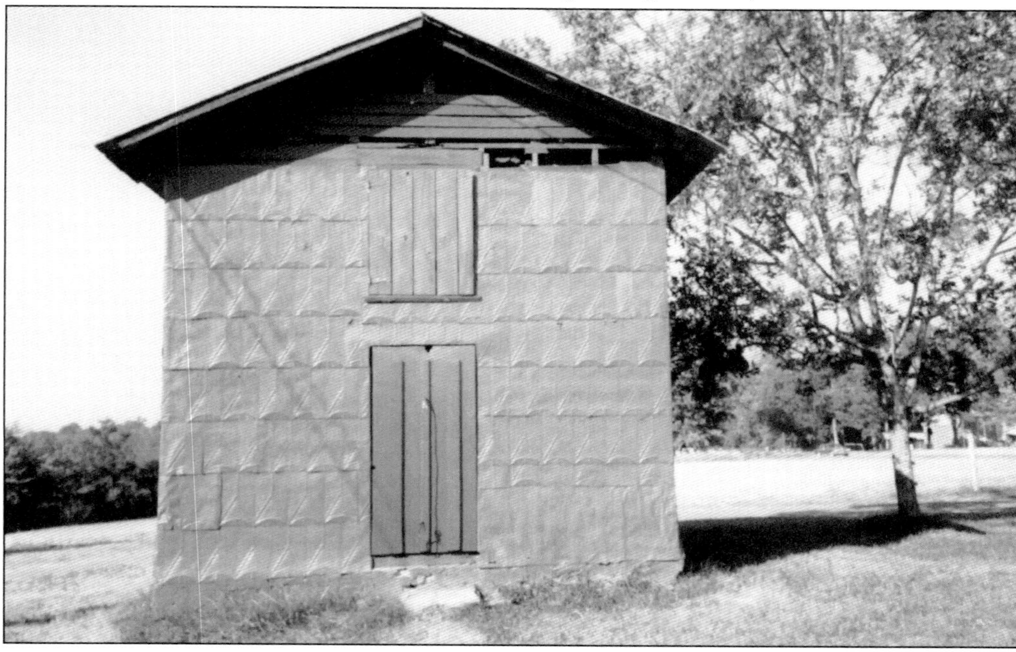

A common sight, this kind of barn was used to smoke meats and was referred to as "the old smokehouse." In its loft, evidence of dripping meats was still apparent in the late 1980s. Being covered in stamped tin gave it an air of style. Most families were somewhat self-sustaining on their farms. People produced their vegetables, fruit, meat, and dairy. (Courtesy of Carol Gilley.)

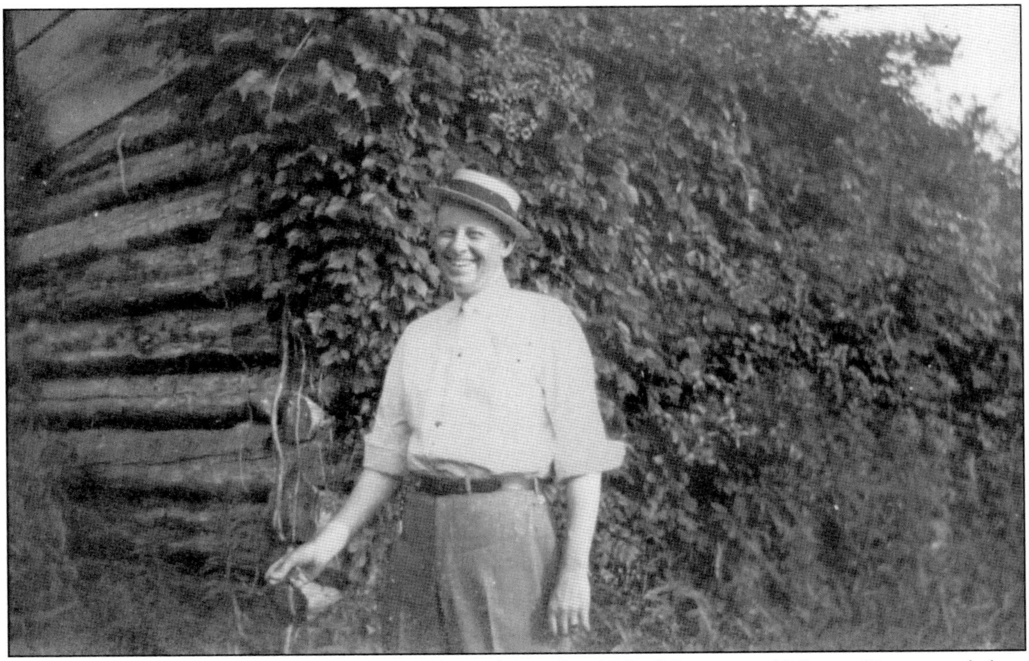

Edgar Babb stands in front of his corncrib in the early 1900s. He married Nancy Scott, and they farmed the Scott home place along Augusta Road. Corncribs stored ears of dried corn, and the boards were separated so air could circulate through the corn to keep it from rotting. Of course, many critters enjoyed the easy access to their favorite dinner. (Courtesy of Helen Babb Campbell.)

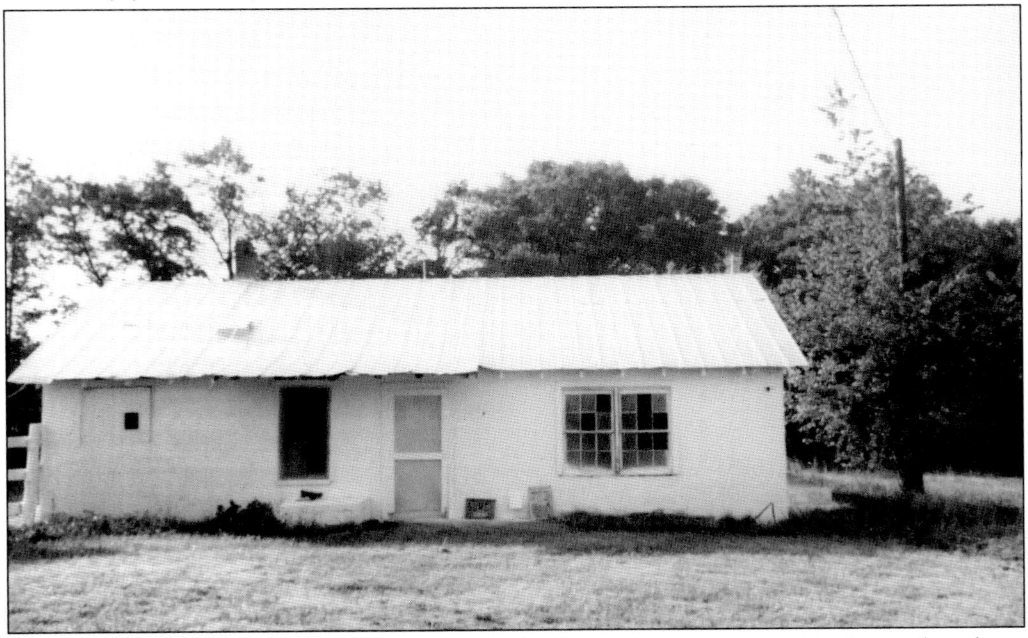

This small dairy barn was in operation until the 1960s and is on the former Jim Terry Place farm. The concrete-block construction and concrete floor made it cool in the summer and easier to clean. Cows were directed to enter at one end, hand milked at four stations, and then ushered out the exit ramp. In the front, the milk was processed and prepared for bottling. (Courtesy of Carol Gilley.)

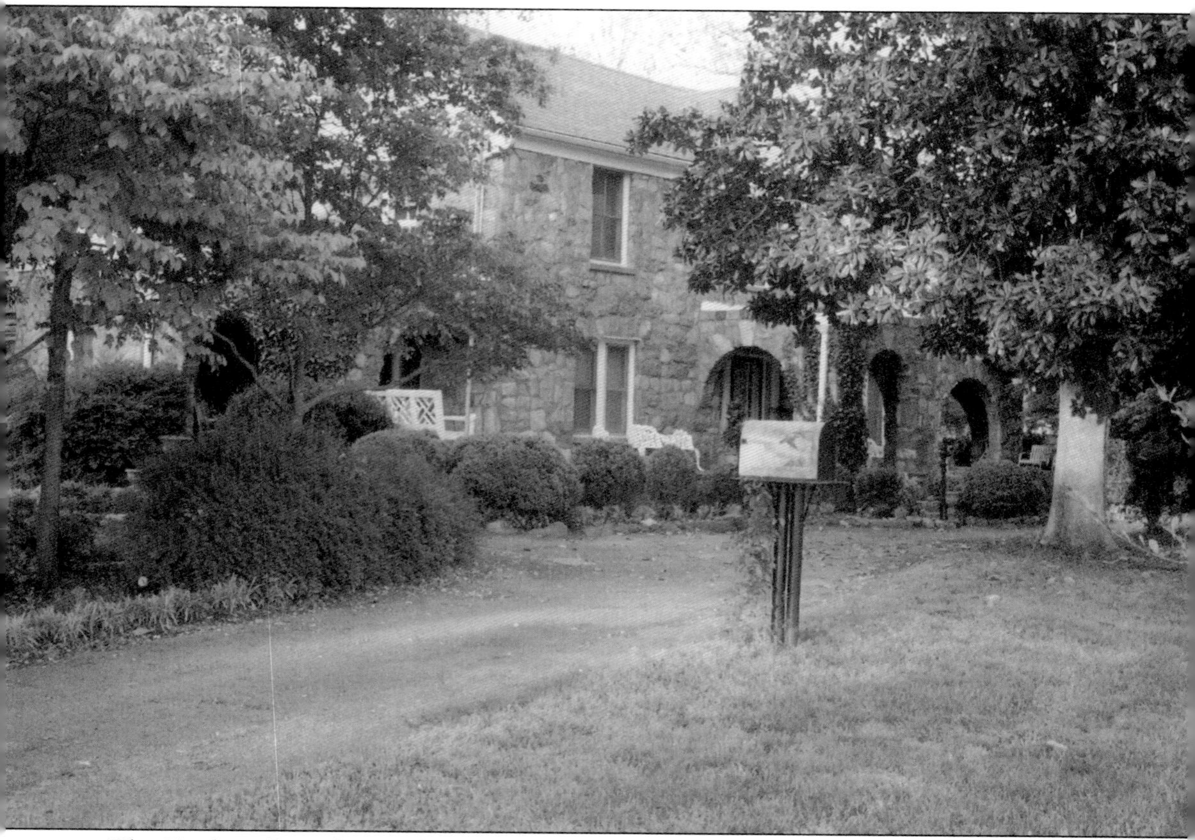

The Hopkins Farm (c. 1840) consists of approximately 340 acres located in the Fork Shoals community in southern Greenville County. Once a plantation of nearly 2,000 acres, it was recognized in the National Register of Historical Places in 2007. The Hopkins complex includes the main house, the cook's house, 11 outbuildings, agricultural fields, a historic pecan grove, and a family cemetery. Under the management of John Drayton Hopkins, the farm came to be a model of innovative farming methods during the Great Depression through the World War II era. A Native American trail marker was discovered on the property. Silk was produced on this farm; threads were spun and woven from silk worms raised on the plantation. During the 1800s, the Hopkinses produced silk, spun it into yarn, and crafted stockings for the governor of South Carolina. (Courtesy of John Drayton Hopkins's Family.)

Six

MAKING CONNECTIONS

COMMUNITIES

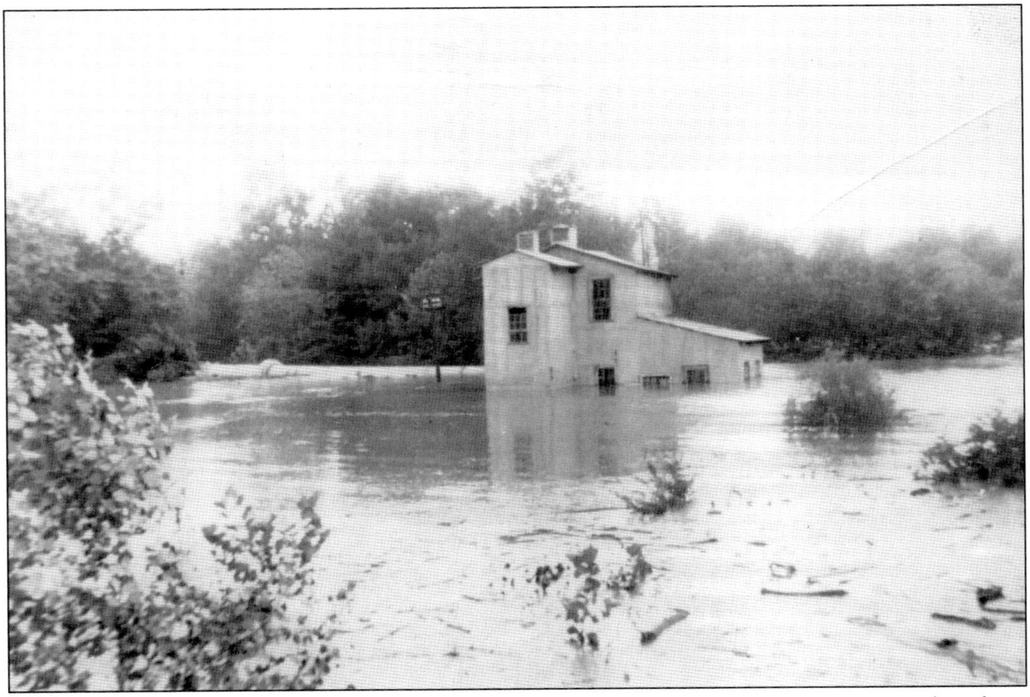

This high water was at Fork Shoals and Cedar Falls in August 1940. This picture was taken from the lower side of the confluence of Huff Creek and the Reedy River. There have been numerous flash floods, but a major flood, the freshet, destroyed everything on or near the river in 1908. Built after 1908, this building is the powerhouse, which contained the hydroelectric generator. (Courtesy of Janice Smith Woods.)

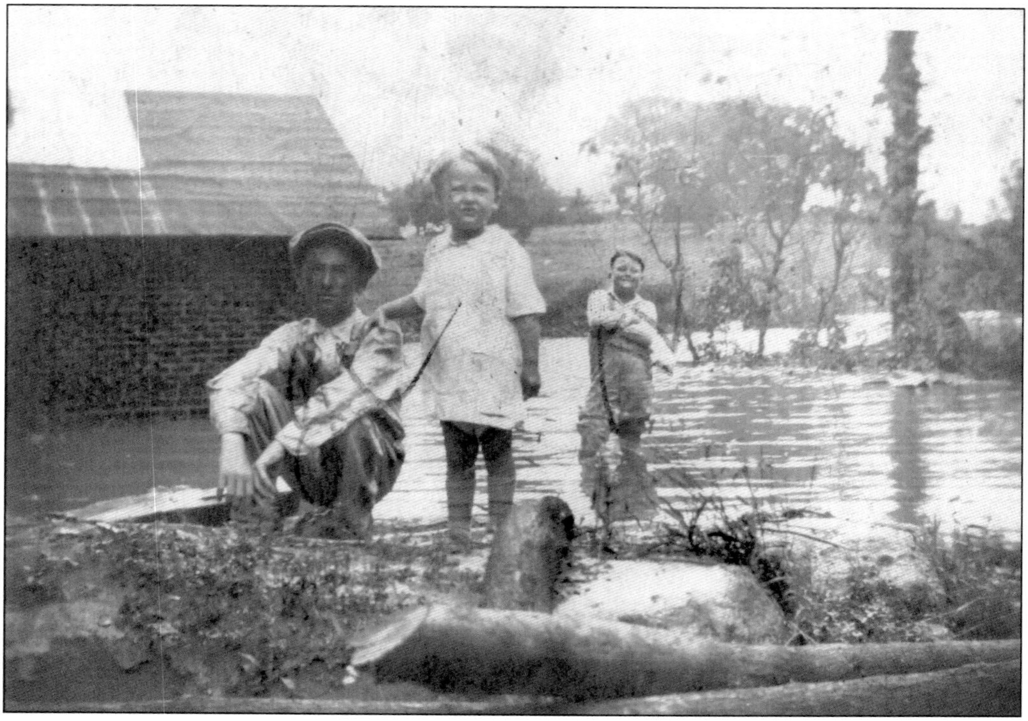

Three boys wade in the floodwaters of Reedy River and Huff Creek near the mill at Fork Shoals. (Courtesy of Gloria Hughey.)

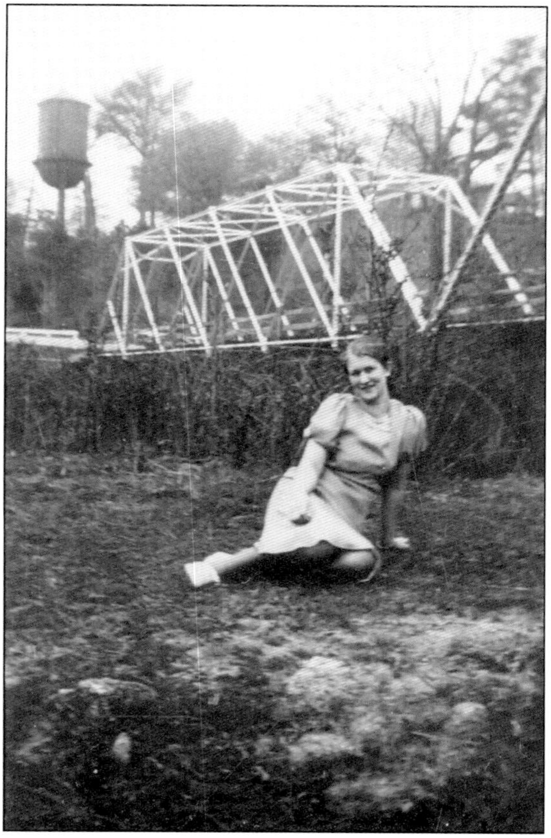

Nannie Harvey Meeks posed for a photograph near the iron bridge beside the Fork Shoals Mill in 1935. (Courtesy of Gloria Hughey.)

Woodrow Wilson, James Calvin Williams, and LaVerne Wilson Williams (pictured from left to right) cooled off at Cedar Falls in the summer of 1948. (Courtesy Dr. Marshall Williams.)

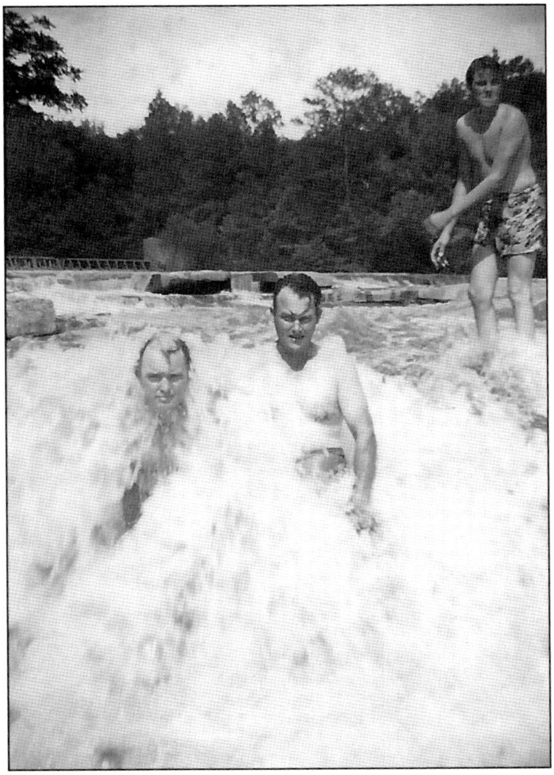

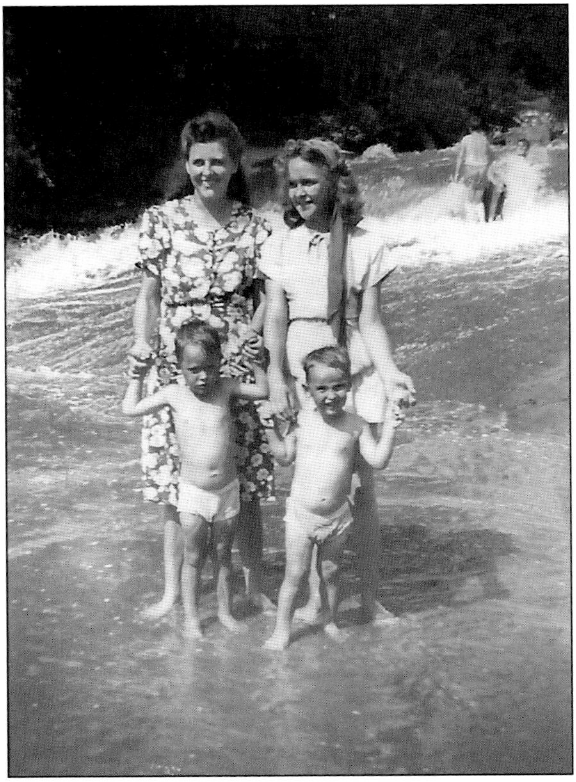

Many happy memories occurred at Cedar Falls, a fun place to cool off on a hot summer day. This photograph was taken during the summer of 1948. The ladies were Louise Lyda Williams (left) and Barbra Joan Gambrell. The boys were James Gerald Williams (left) and Doyle Fulton Williams. (Courtesy of Dr. Marshall Williams.)

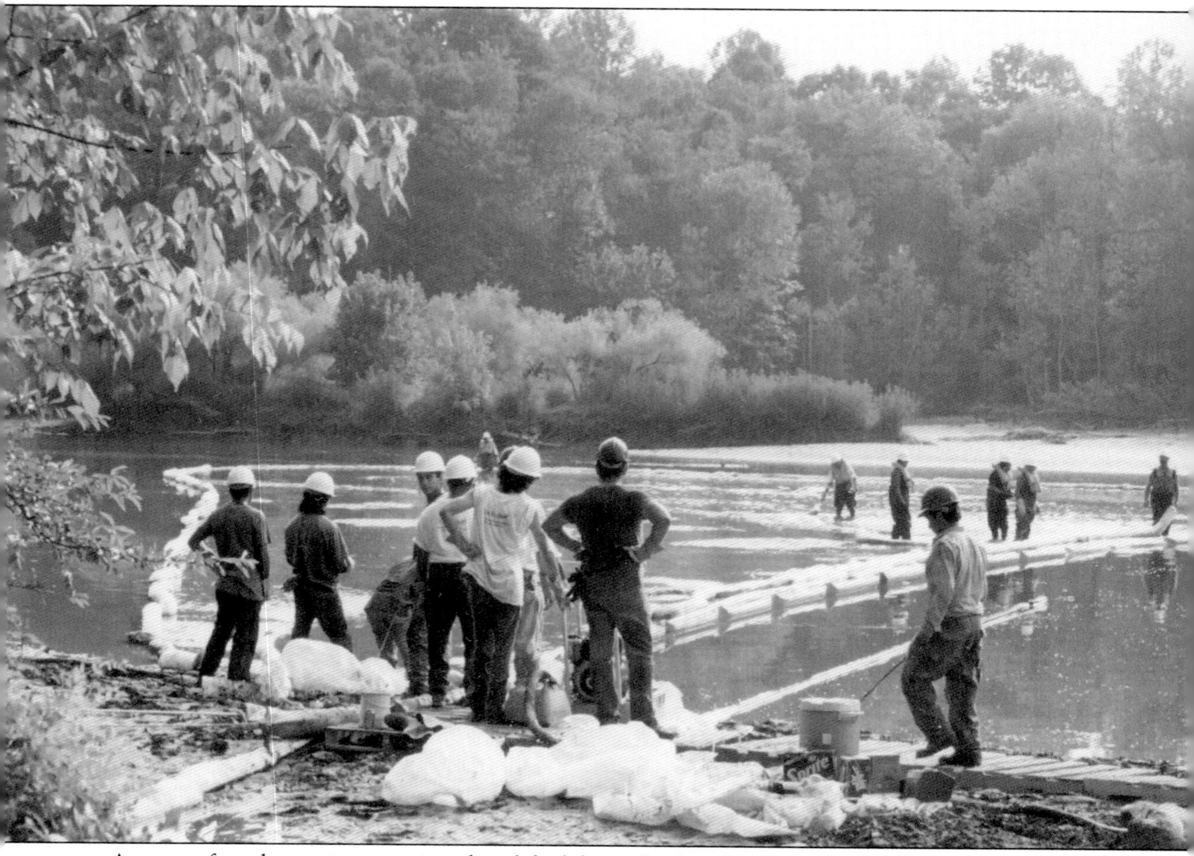

A crew of workmen is removing diesel fuel from the Reedy River near Cedar Falls. In 1996, an oil spill by Colonial Pipeline dumped thousands of gallons of diesel fuel into the Reedy River just north of Fork Shoals community. A massive clean-up effort followed to return the river to its environmental condition before the spill. (Courtesy of Canebrake Fire Department.)

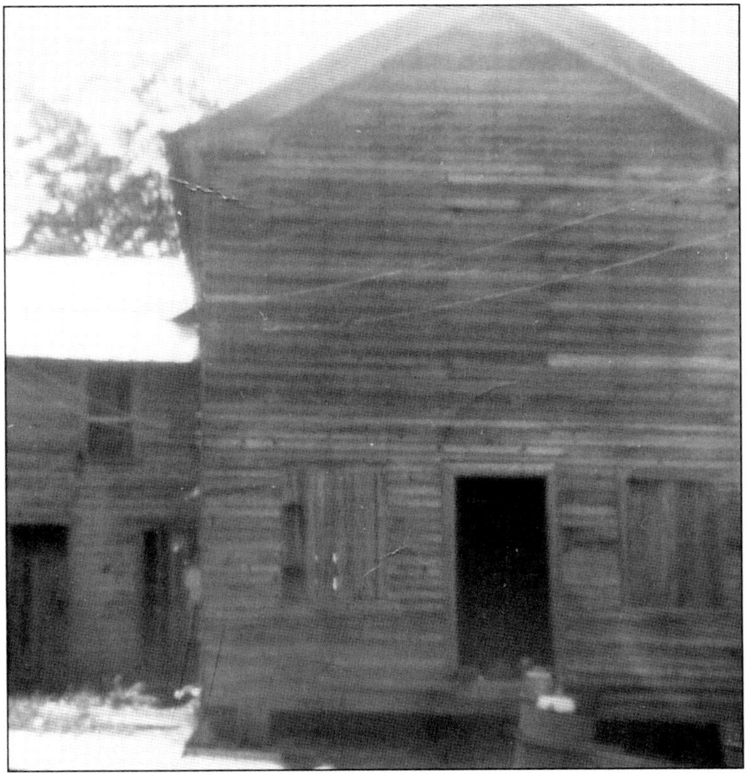

Fork Shoals's first school building was erected on the grounds of Fork Shoals Baptist Church. This is the only known photograph. Constructed in 1886, the building was erected by the Masons of Ornan Lodge, which was located at Cedar Falls prior to this building. The downstairs was used as the school, and the masons used the upstairs for their meetings. (Courtesy of Lou Ann Baldwin.)

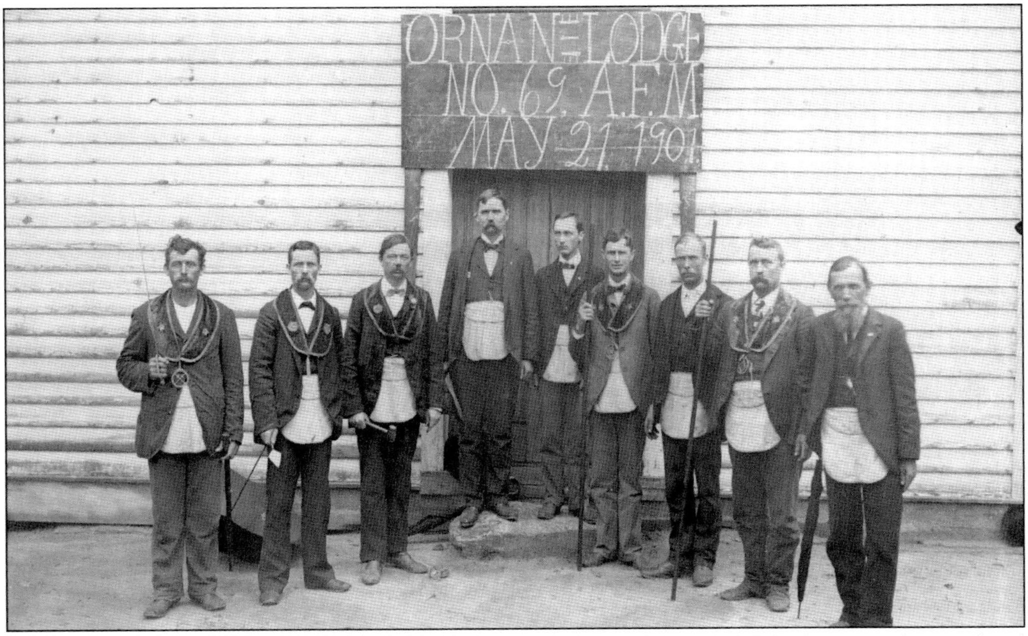

On May 21, 1901, members stood in front of the Ornan Lodge Hall, which also served as a school in Fork Shoals. This was most likely a celebration of the lodge's 50th year. This building was demolished in 1949, and a new one was built on the same property in celebration of the lodge's 100th birthday. A 163-year history makes Ornan the oldest organized lodge in Greenville County. (Courtesy of Jim Scott.)

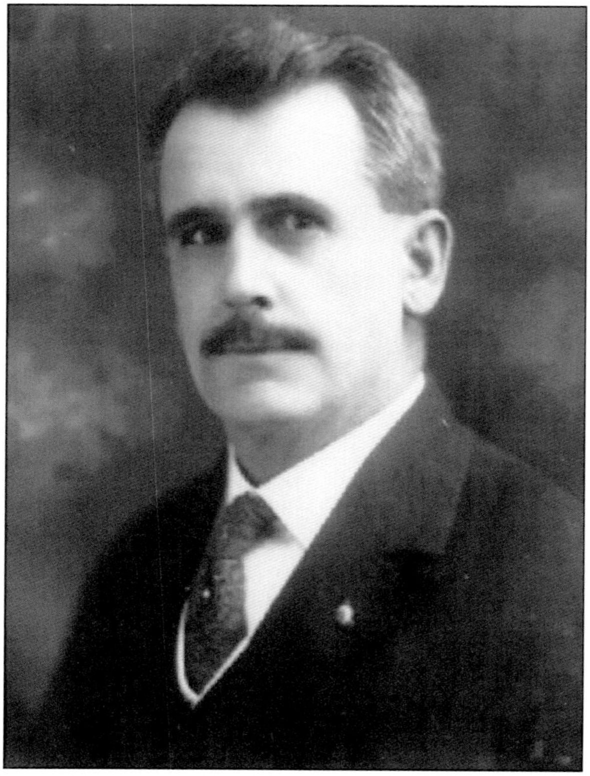

Gov. Robert Archer Cooper (1874–1953) was a member of Ornan Lodge in Fork Shoals where he lived after moving from the Waterloo Township of Laurens County. Cooper was governor of South Carolina from 1919 to 1922, when he resigned to become a member of the Federal Farm Loan Board. He also served for over a decade as US district judge for Puerto Rico. (Courtesy of SCIWAY.)

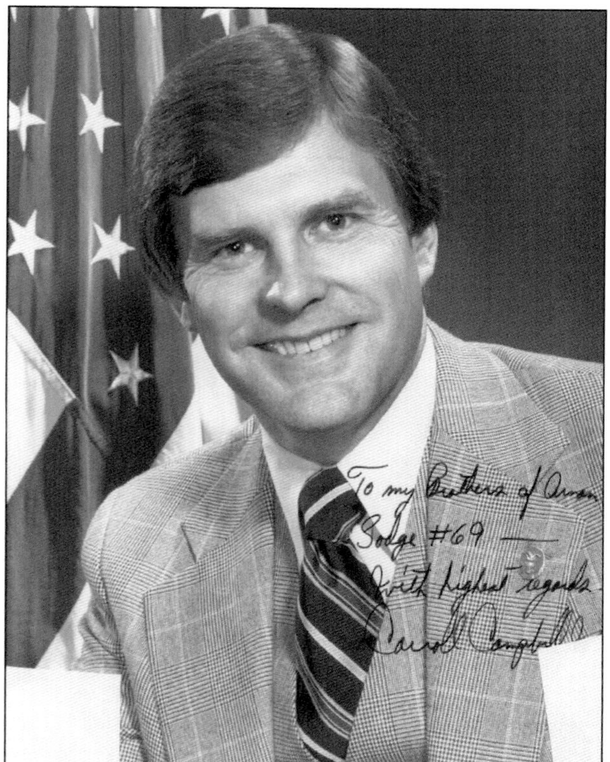

Gov. Carroll Ashmore Campbell Jr. (1940–2005) became a distinguished member of Ornan Lodge in 1978. After serving in the House of Representatives and Senate in the state, he served in the US House of Representatives from 1970 to 1987, when he was elected governor of South Carolina. He served in that office from 1988 through 1992. (Courtesy of Jim Scott.)

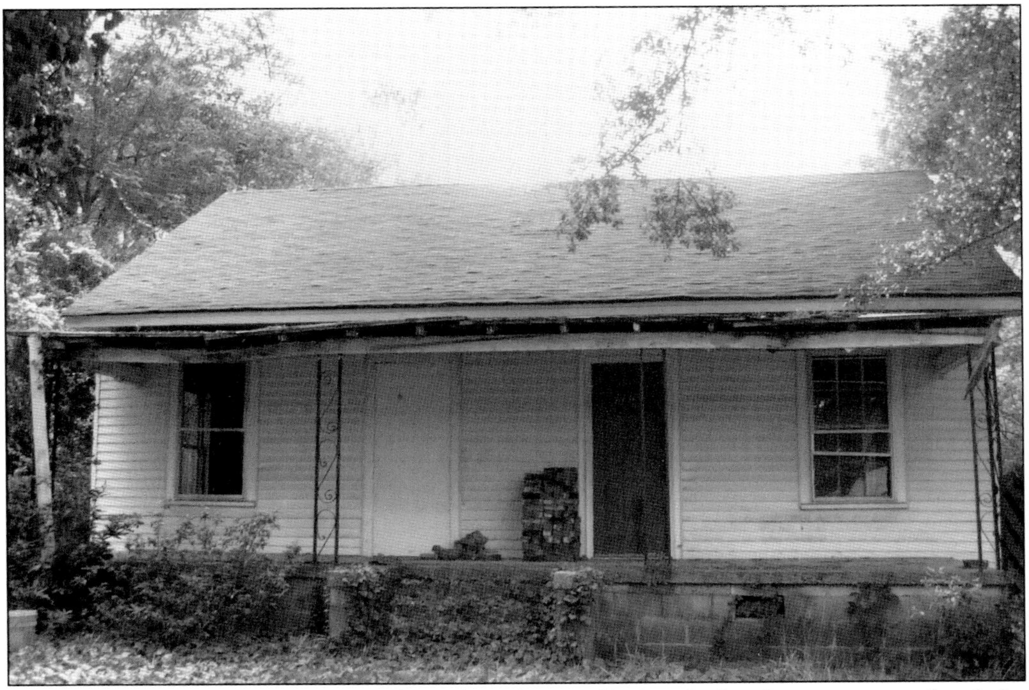

Here is a typical mill house in the days when Greenville, South Carolina, was considered the "Textile Center of the World." L. Harvey owned this house in Fork Shoals, and it was here where Jr. (1931) and Janice (1934) were born. This photograph was taken in 2012 by Carol Gilley. (Courtesy of Carol Gilley.)

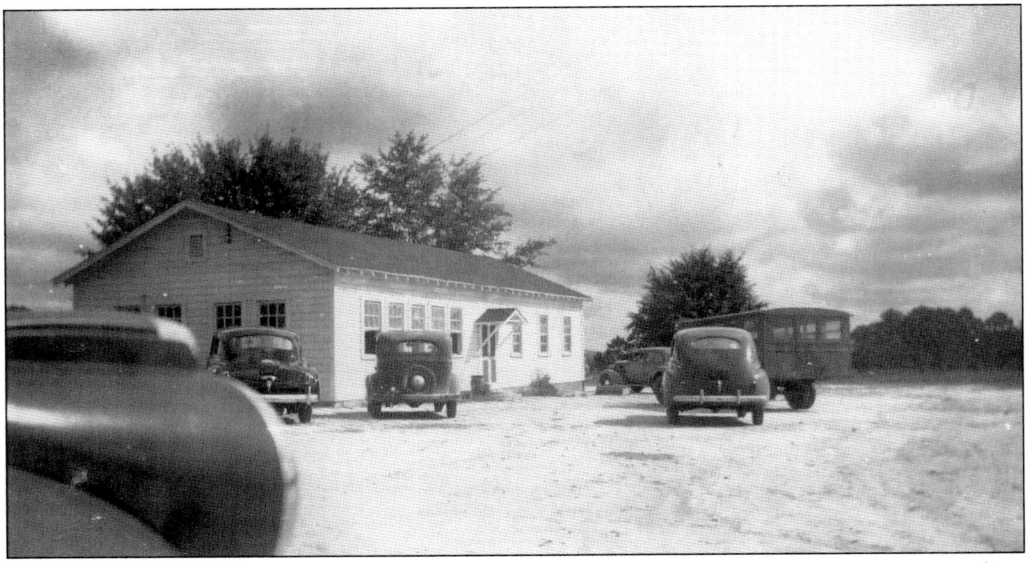

The Fork Shoals cannery was a busy place. The cans were often marked with a letter as to their contents with the scratch of a large nail. Sometimes supper was a surprise because "P" stood for peaches or for pears. (Courtesy of Cheryl MacKnight.)

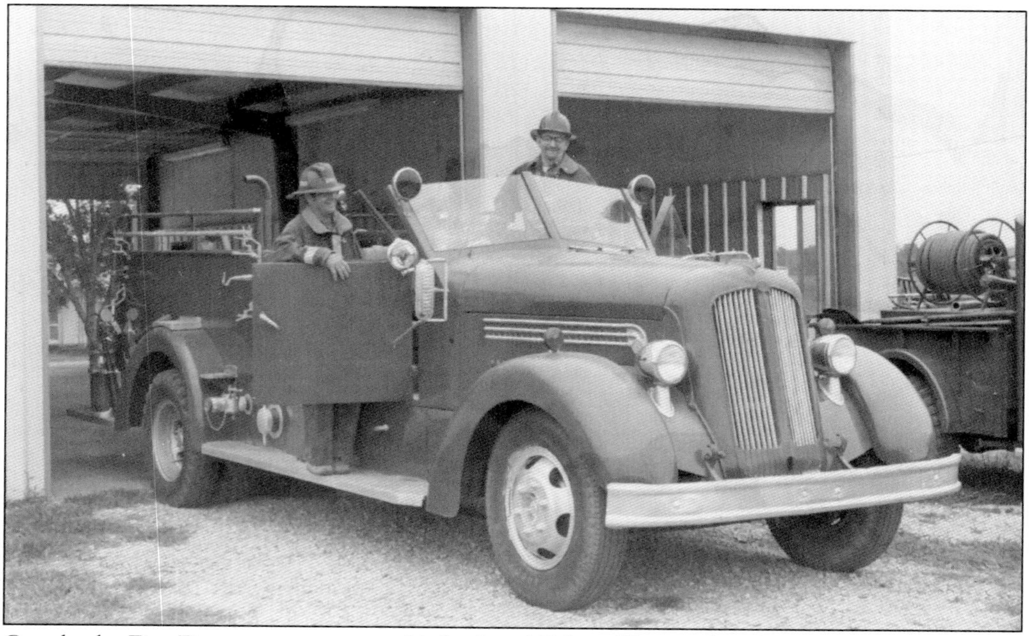

Canebrake Fire Department was established in 1985 to fill a void in fire protection service in Fork Shoals and the surrounding area. Predominately a volunteer fire department, it has a very good reputation for the work done. This 1954 fire truck was bought to save money as a fledgling department. The headquarters is located on McKelvey Road, a few miles east of Fork Shoals. (Courtesy of Canebrake Fire Department.)

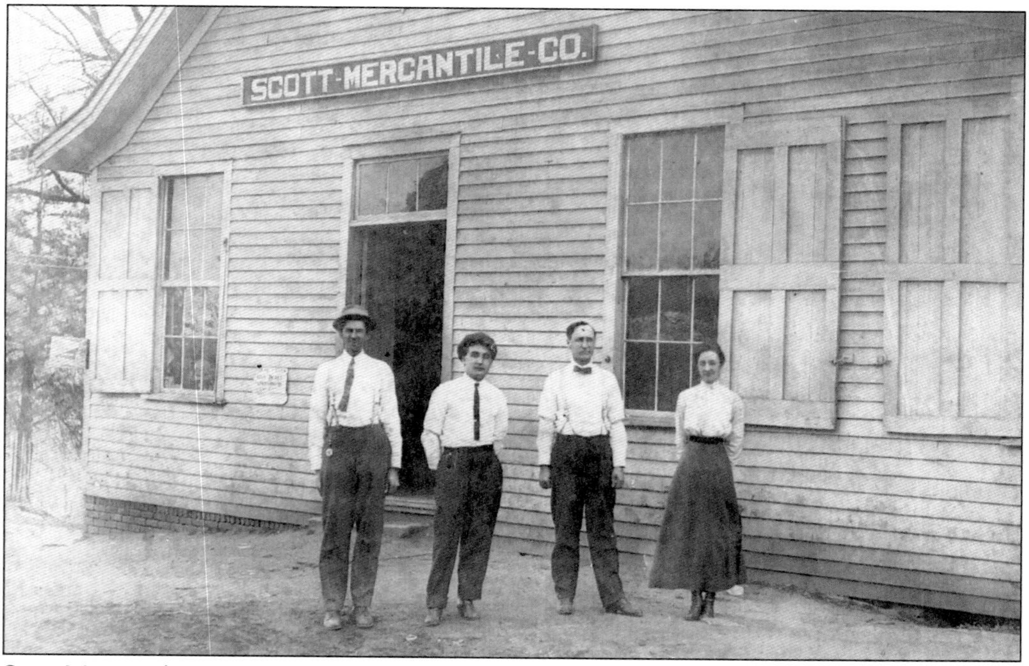

Scott Mercantile Company stood near the current-day Canebrake Fire Department Substation on McKelvey Road in Fork Shoals. Shown in the picture are, from the left, Ped Kellett, Alex Rogers, Bob Scott (store proprietor), and Lucy Sprouse. Fork Shoals was a thriving community in those days around the turn of the 20th century. (Courtesy of Scipio Peden and Lewis Terry.)

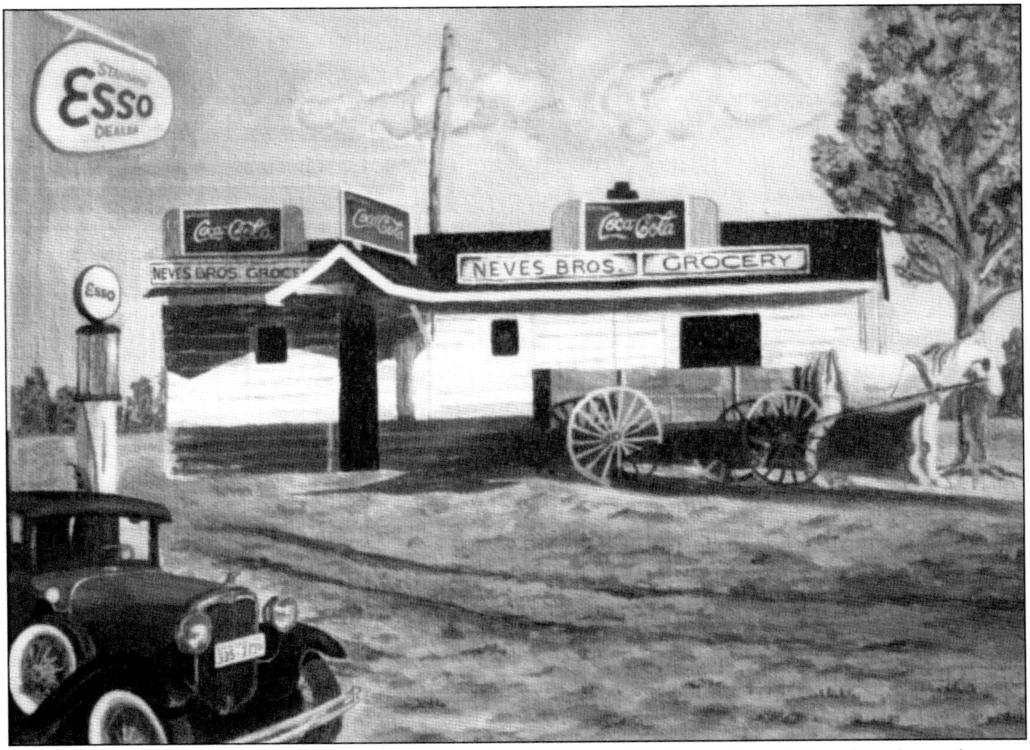

Neves Bros. Grocery was located at the corner of McKelvey and Hillside Church Roads, a few miles east of Fork Shoals. This was a thriving community, and the store continued until the late 20th century. Note the two types of transportation—horse and wagon and the automobile. The headquarters of Canebrake Fire Department now occupies this corner. (Courtesy of the Neves family.)

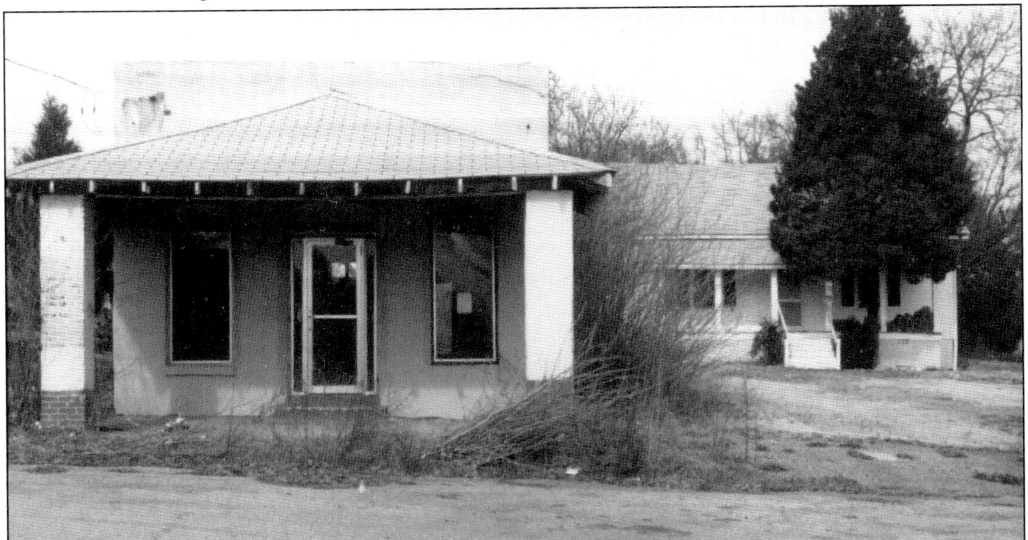

The photograph of the Old Bank Building in Fork Shoals was taken in 1984. The building became a general store under various proprietors after the bank vacated the building. It stood at the intersection of McKelvey, Berry, and Cedar Falls Roads in Fork Shoals. Red Hill and the Hipps family were proprietors in the 1940s through the 1950s. (Courtesy of Doyle Williams.)

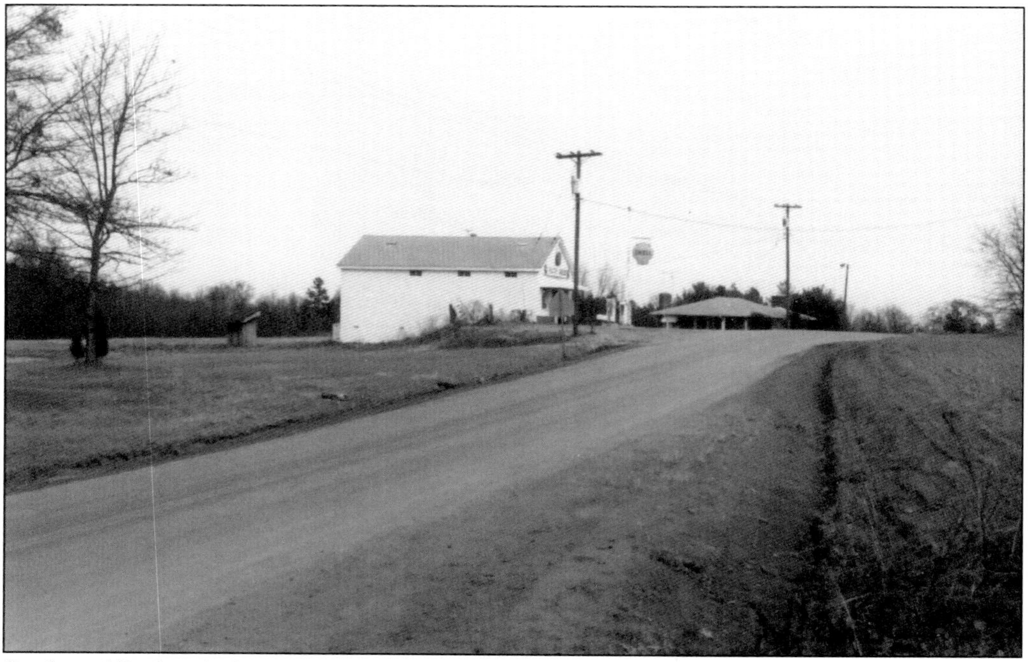

Gault and Rodgers's Grocery Store is pictured in the early 1980s. This store was directly across the road from the old Fork Shoals High School and in front of Fork Shoals Baptist Church. Frank Gault married Mr. Rodgers's daughter Wilma. This store served the Fork Shoals community for many years and was the last store to operate in Fork Shoals. Frank and Wilma's house is to the right of the store, and Wilma still lives there. (Courtesy of Doyle Williams.)

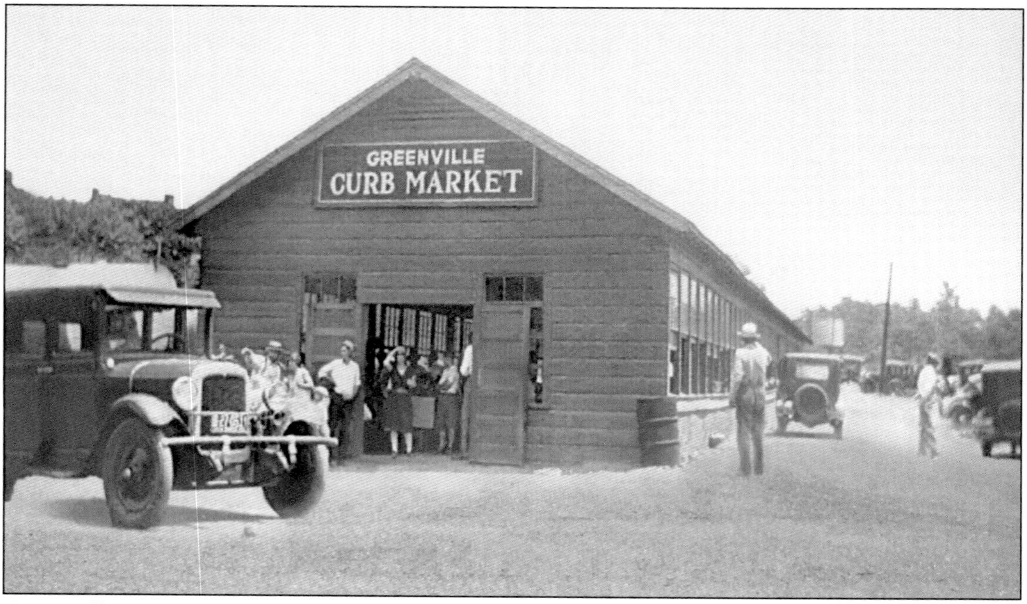

Greenville Curb Market in downtown Greenville was a good place for Fork Shoals area farmers to peddle their farm products. They sold produce, beef, pork, chickens (live and processed), milk, eggs, butter, and other farm products. The ladies baked cakes and pies to sell and also had quilts and other handmade articles available for purchase. It was a cooperative effort between the city and the farms that benefited both. (Courtesy of Cheryl MacKnight.)

In the 1950s, Bruce Sims (right) raced this competition coupe. In 1959, Sims bought a railcar and won the SC Championship around 1961. It is now in Garlit's Drag Racing Museum in Oscala, Florida. Jean Ware Toole operated the Esso Station in this picture for many years. This building still stands and is used as an automotive shop. (Courtesy of John Sims.)

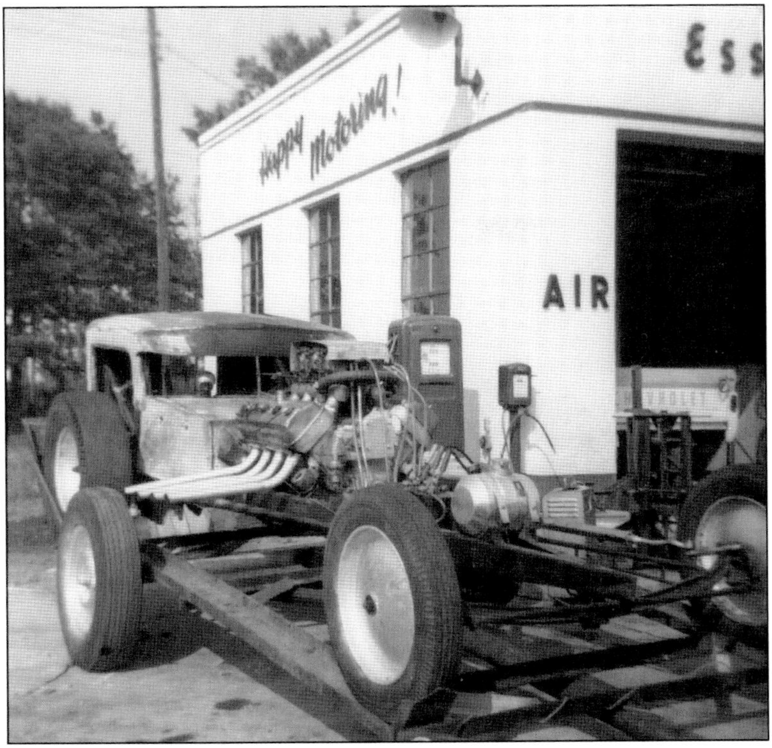

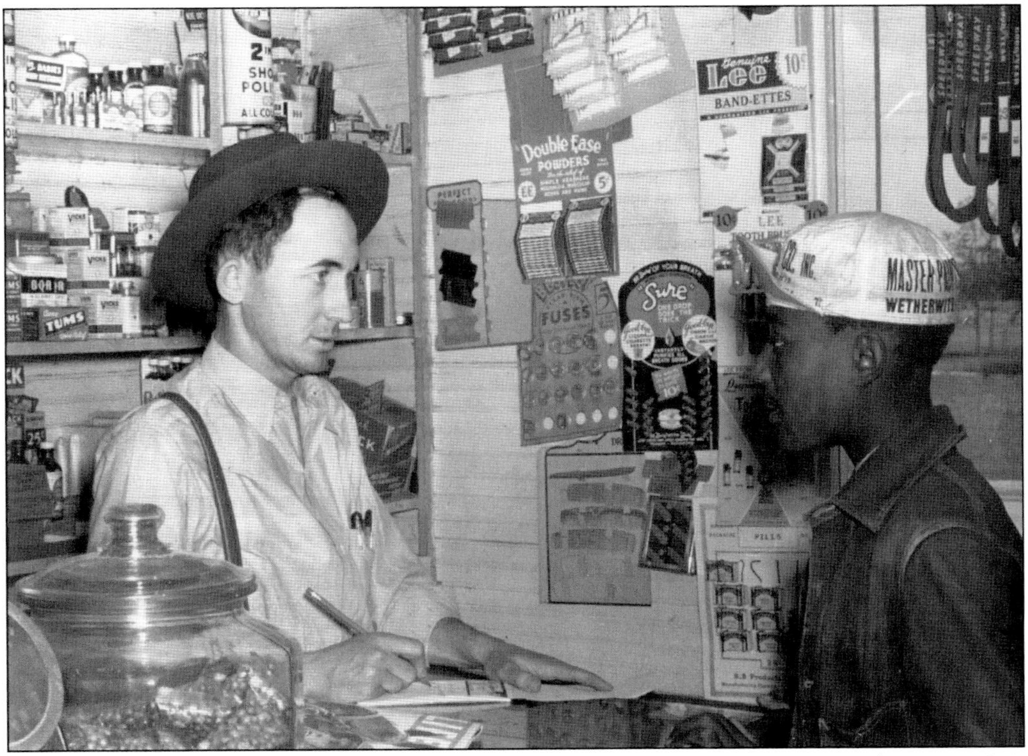

J.R. Chandler's store was a key location at the Ware Place. J.R. and a costumer are shown in 1941. The general store also carried medicines and car parts. (Courtesy of J.R. Chandler IV.)

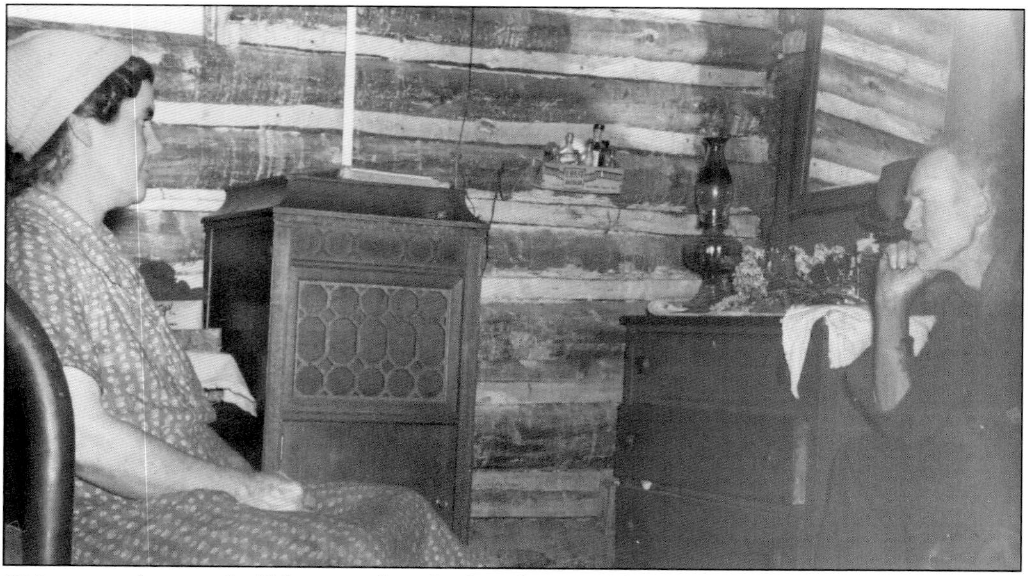

Citizens rarely turned off their speakers for fear they would miss something from the Chandlers' grapevine radio broadcast. News of any kind in the community that anyone wanted to share would be reported to the Chandlers so they could announce it via the wire, be it a death, an emergency, or that somebody's cow was out. There was a system in place to tie in any chapel program or special programs with the schools. Mostly for entertainment, speakers were almost always left on. A piano and pulpit in the studio enabled anyone, with any good talent, to broadcast themselves to an audience of a few thousand listeners. (Courtesy of J.R. Chandler IV.)

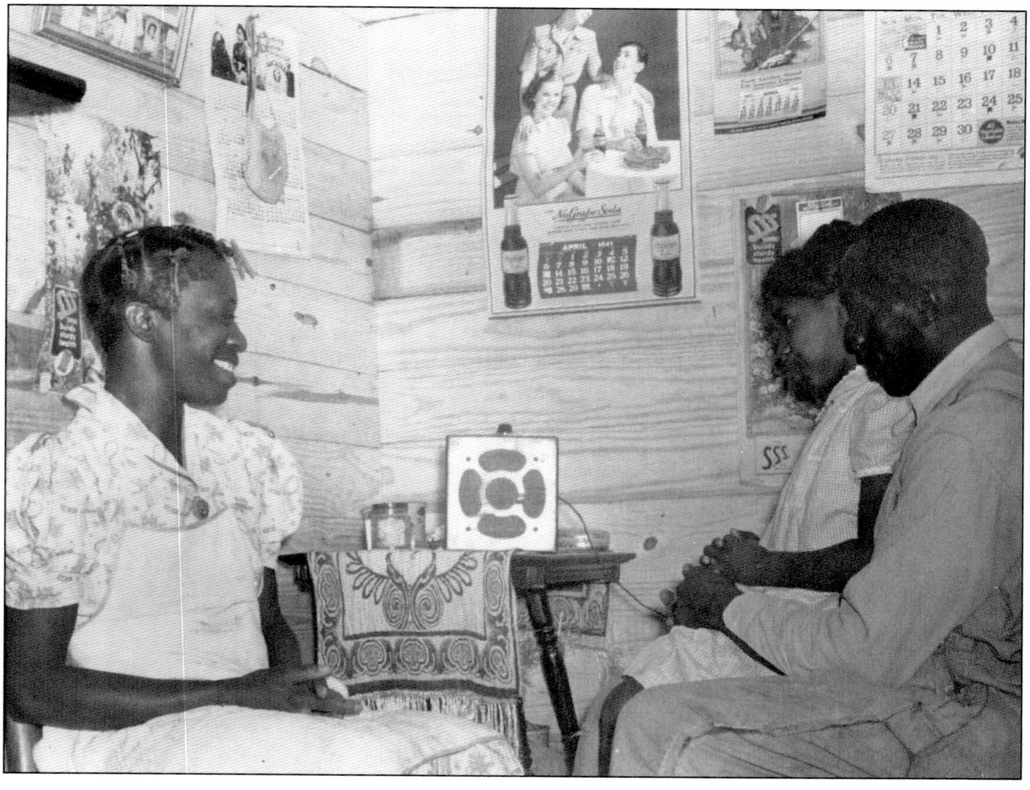

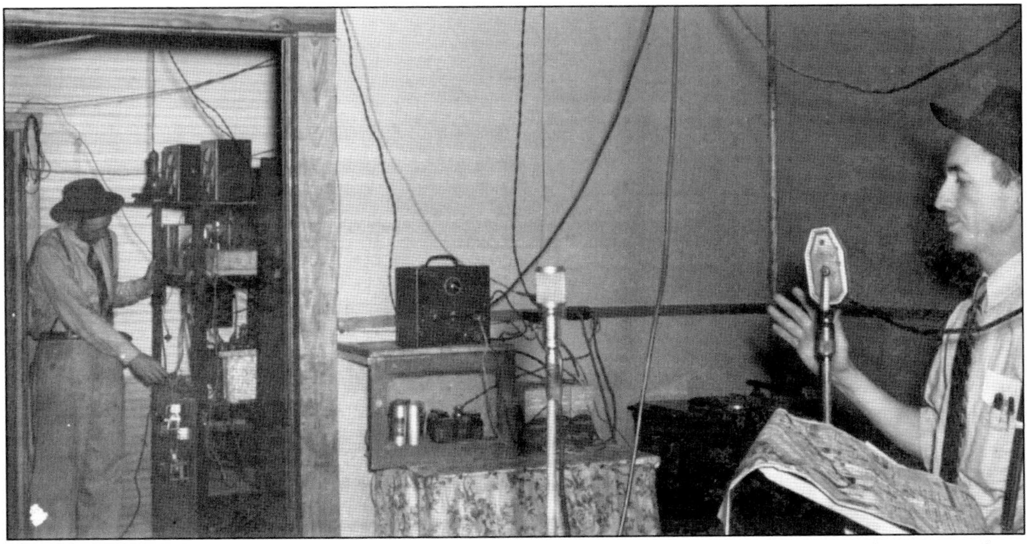

J.R. Chandler Jr. (right) and his cousin J.E. Chandler operated a general store at the Ware Place from the 1930s until the mid-1970s. They planned and built a network of speakers connected to one huge radio and amplifier. Gordon Rogers of Mauldin, a Clemson student, who was assisted by Sherman Fox, built the amplifier. They broadcasted from the studio at the rear of their store through a network of wires connected to 1,000 homes up to about four miles away in all directions. Anyone could "get hooked on" for 35¢ a month and have perpetual radio service. It served as a one-way communication device for news and local talent. Sometimes, they placed the microphone in front of a radio to relay national programs. (Courtesy of J.R. Chandler IV.)

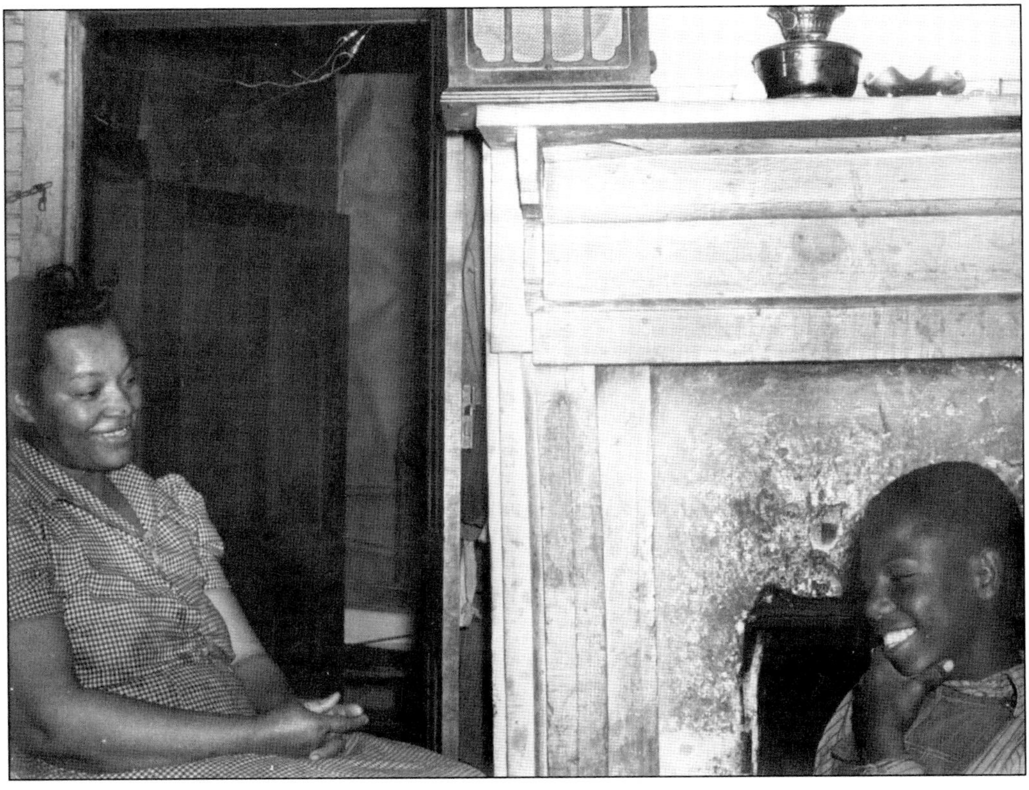

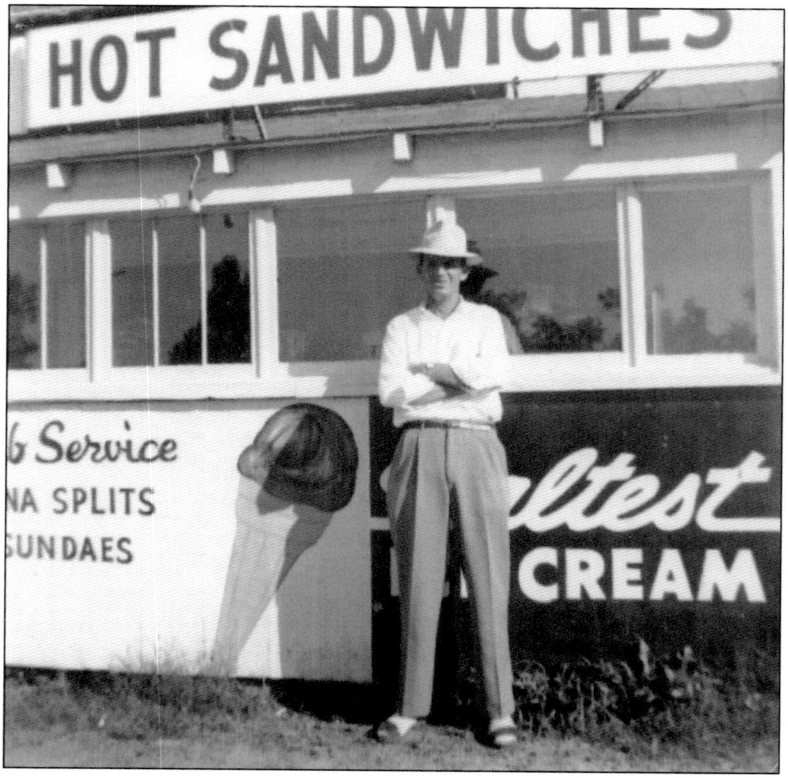

Harley Moon is standing before the infamous Grill, which he opened in a bus in 1947 at Ware Place. When the bus burned in 1948, Moon rebuilt on the same site. The Grill was frequented daily by teenagers from Ellen Woodside High School for their delicious hot dogs, hamburgers, and cherry cokes. In the 1950s, a skating rink was added. It burned in 1962 or 1963. (Courtesy of George and Elaine Moon.)

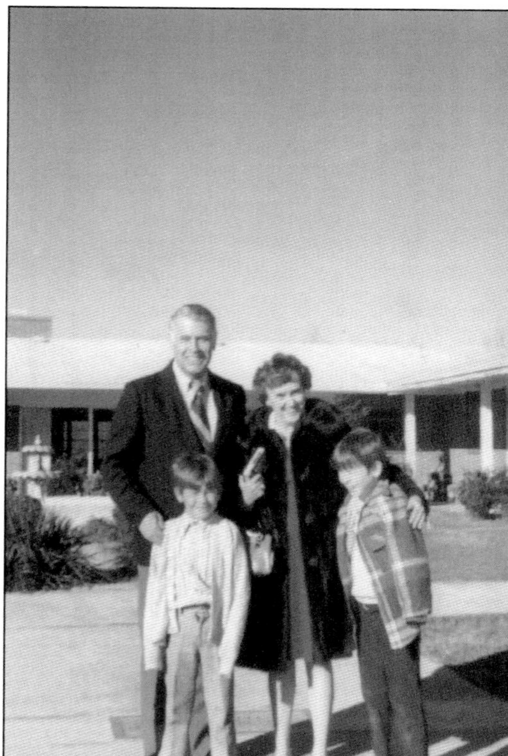

Charles Aiken, founder of Boys' Home of the South in 1957, is seen here with a guest and two resident boys. Aiken dedicated his work to caring for orphaned boys and over time has provided care for hundreds of children from age 6 to 21. His mission was to "Mend Boys and Build Men." (Courtesy of Debbie Pitts Moorer.)

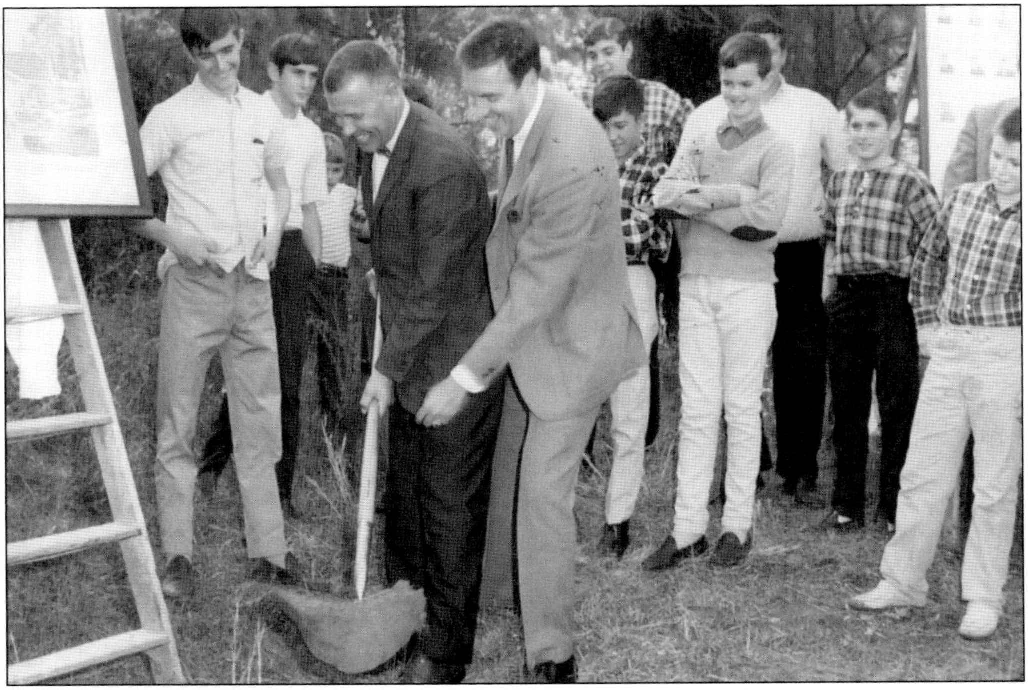

Charles Aiken and Jim Nabors (Gomer Pyle of *The Andy Griffith Show*) with lots of the boys break ground for the campus chapel at the Boys' Home on US 25 South below the Ware Place. Nabors donated to the construction of the chapel, which was named for him. (Courtesy of Debbie Pitts Moorer.)

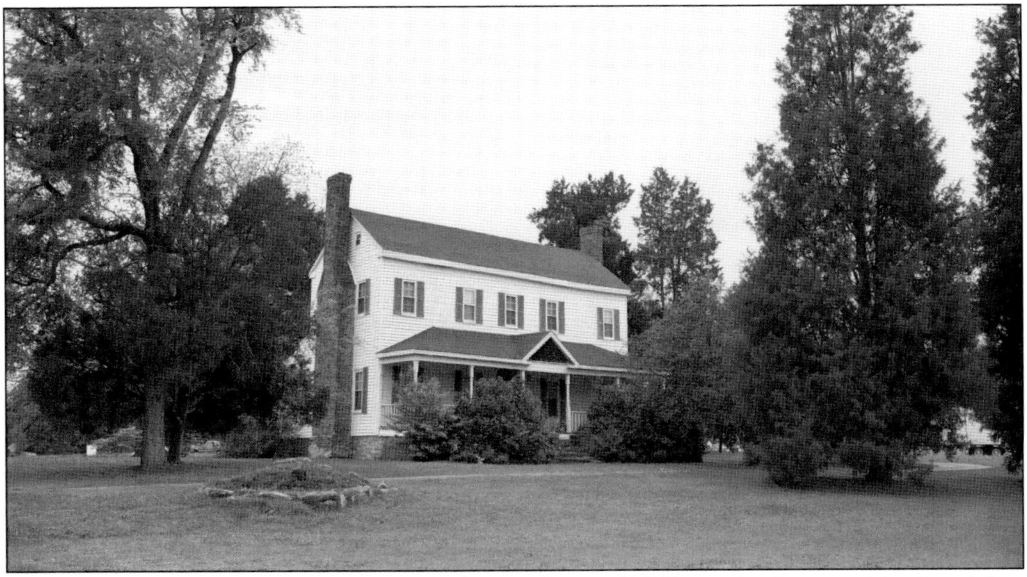

The Ware house, built in the 1830s, was the home of Samuel Williams, but his granddaughter Mary Jones married Thomas Edwin Ware. Ware was a large plantation owner in the early 1800s and served in the South Carolina House of Representatives and Senate. During a family disagreement, Senator Ware shot his father-in-law, Adam Jones. He was convicted of the crime and imprisoned for one week. The governor at the time pardoned him ordering him to pay a hefty fine. (Courtesy of Fork Shoals Historical Society.)

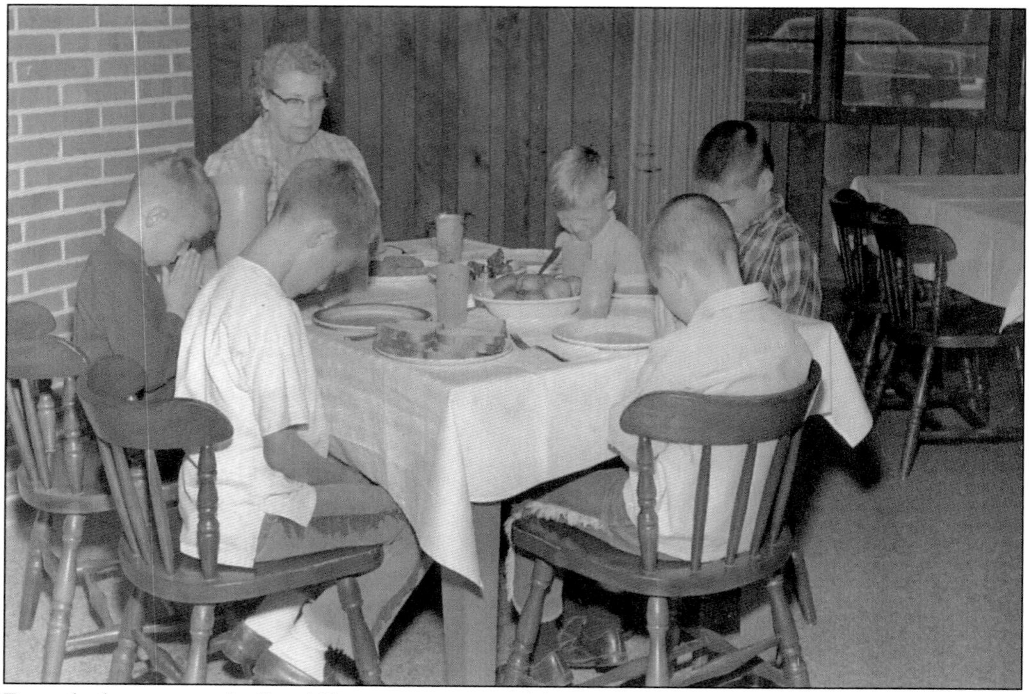

From the beginning, the Boys' Home set out to model positive behaviors for their young folks. These children are participating in a meal as a family. Aiken's wish was to provide opportunities for building young men's strengths, spirits, and spirituality. (Courtesy of Debbie Pitts Moorer.)

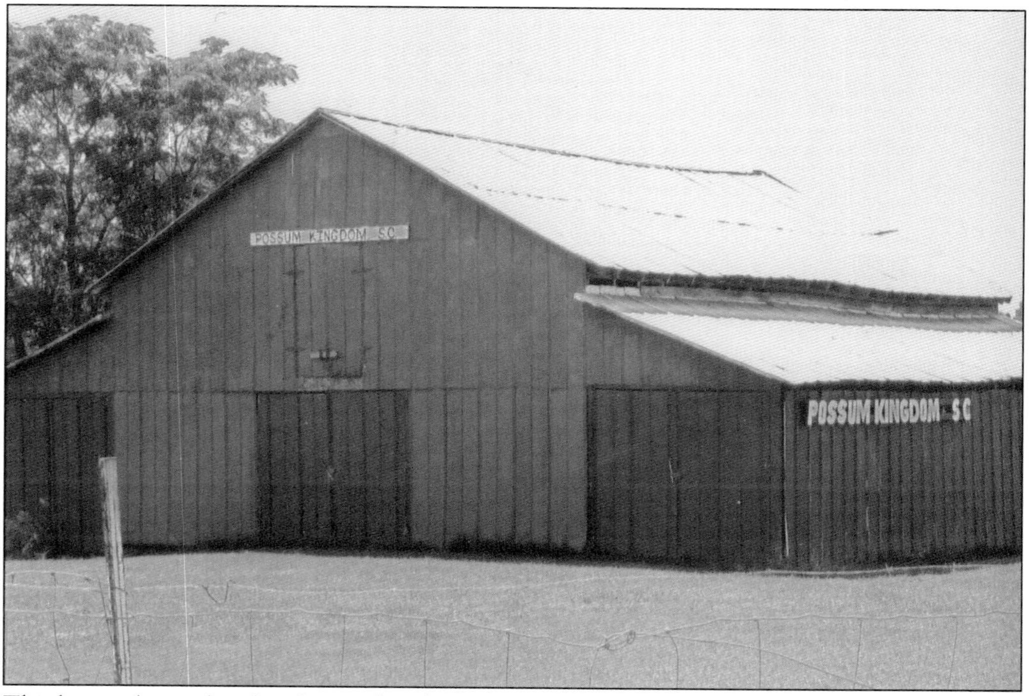

This barn is located in the elusive place known as Possum Kingdom. At one time, no one claimed to be from that area of southwestern Greenville County. Now, to many, it is a claim to fame. (Courtesy of Anne Peden.)

The original Cooley Bridge was built across the Saluda River in 1835–1836, replacing a ferry. Hiram Cooley, a cotton planter, who owned a large plantation and operated a cotton gin and gristmill nearby, built the bridge. This photograph was probably taken around the turn of the 20th century. Note the horse-drawn covered wagon and World War I soldier. (Courtesy of Old Hundred Store.)

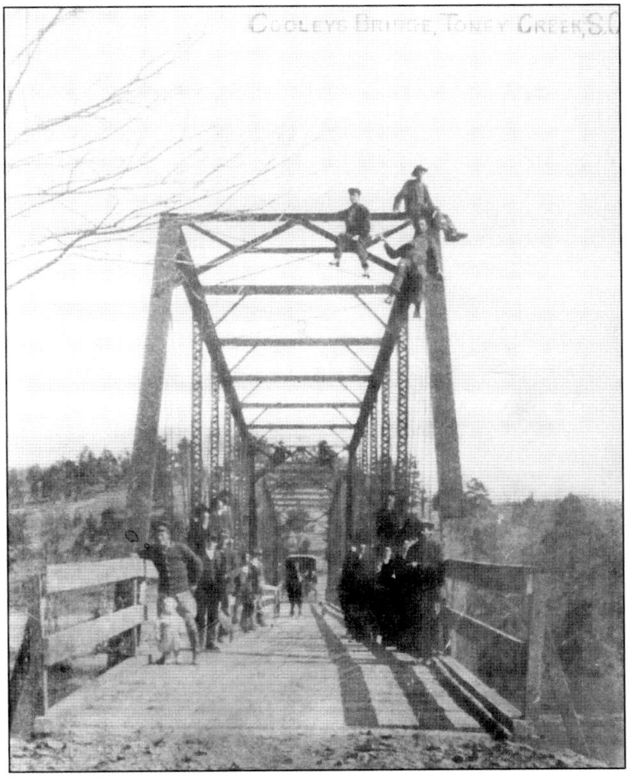

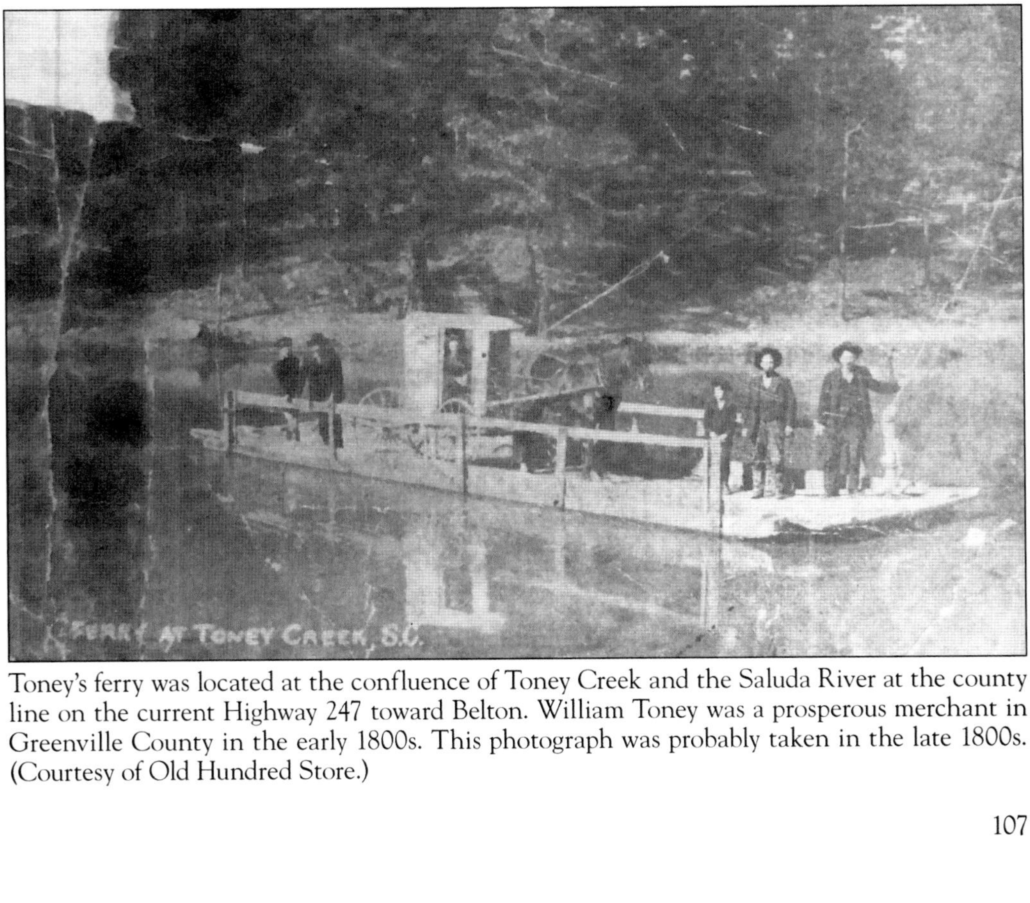

Toney's ferry was located at the confluence of Toney Creek and the Saluda River at the county line on the current Highway 247 toward Belton. William Toney was a prosperous merchant in Greenville County in the early 1800s. This photograph was probably taken in the late 1800s. (Courtesy of Old Hundred Store.)

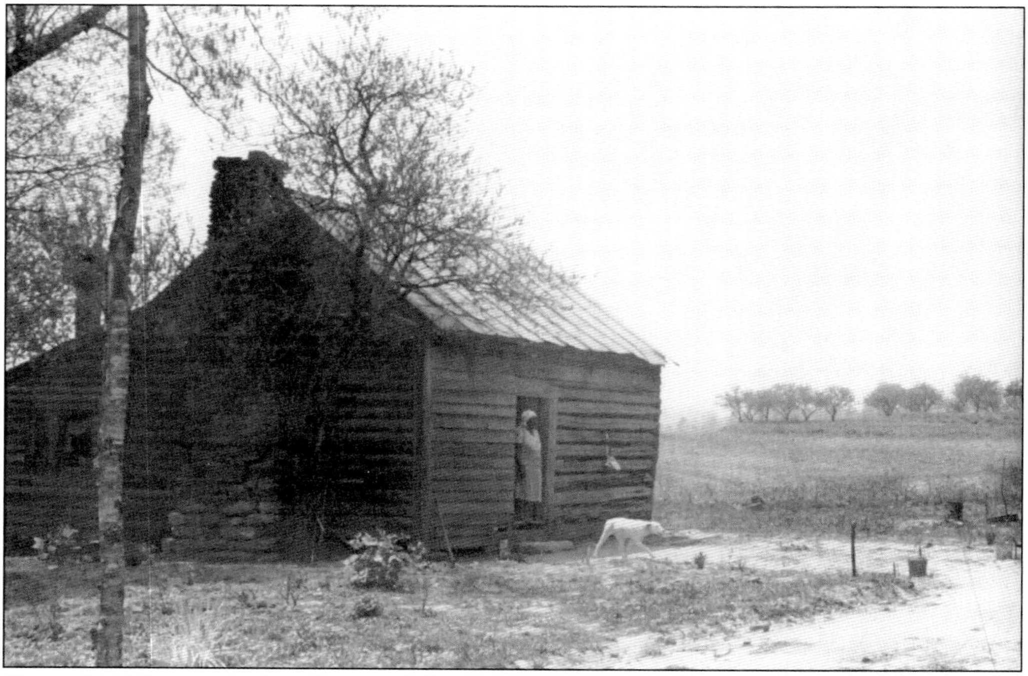

Gertrude Williams, who lived here with her dog a short distance behind the Chandlers' store, was featured in a 1940s Greenville newspaper article about the Chandlers' radio network. The south was changing, and the driving force of the economy continued to shift away from agriculture. Homes, such as hers, photographed in 1941, began to disappear from the landscape. Very few remain standing today. (Courtesy of J.R. Chandler IV.)

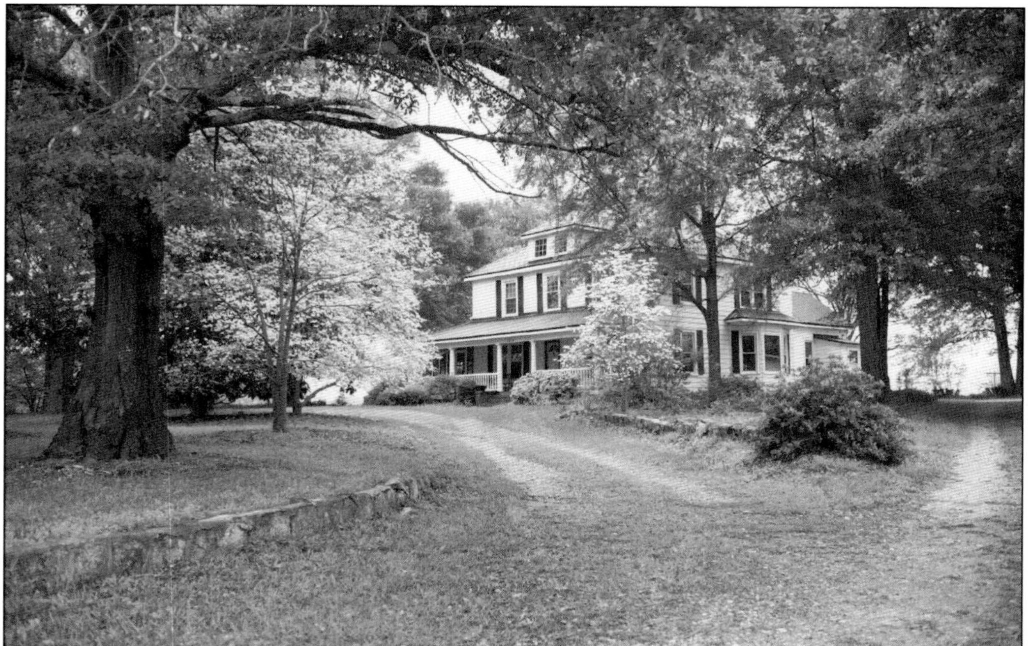

Dr. W.A. Ross, the earliest physician in the area, built this distinctive home in Fork Shoals. It is still a landmark in the village today. (Courtesy of Fork Shoals Historical Society.)

Seven

Caring Souls

Folks

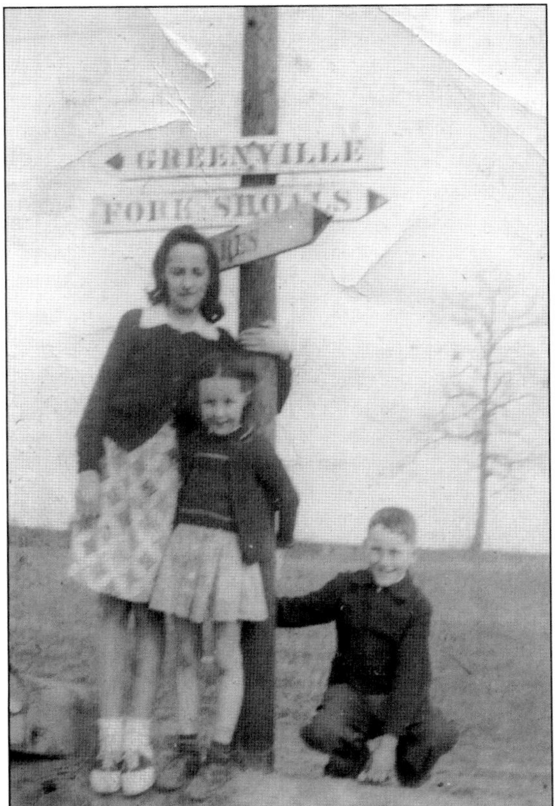

"Which way do we go?" In this 1950s photograph, these young folks, Sara Tumblin Holt, Margie Mae, and Wallace Tumblin (pictured left to right), seem not to care. They are just ready to stay put in Fork Shoals. (Courtesy of Lou Ann Baldwin.)

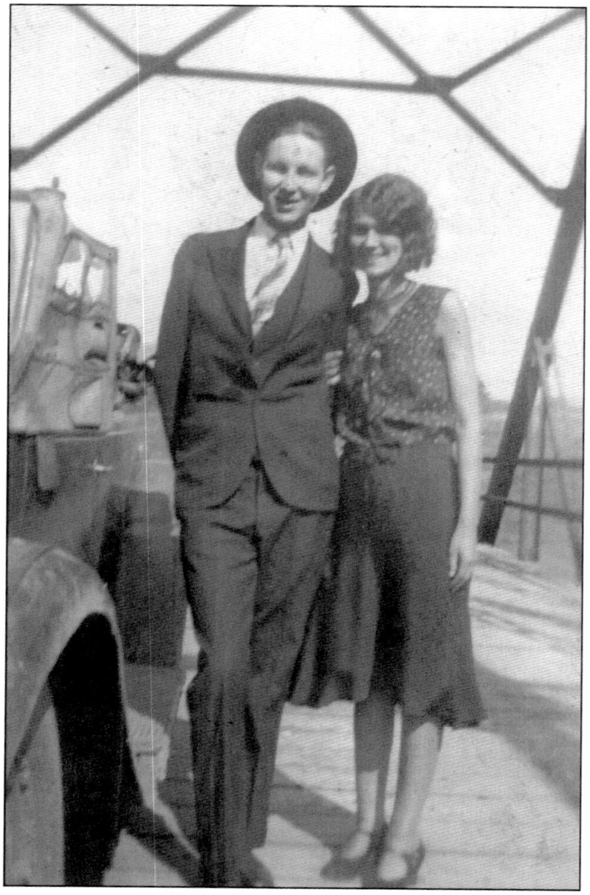

O.B. Harvey and friend enjoyed the day on the old iron bridge over the Reedy River. This was probably taken in the 1930s. The bridge was a favorite place for photographs during that time. (Courtesy of Gloria Hughey.)

Probably taken at the corner of Berry and McKelvey Roads in front of one of the double-door mill houses, this photograph is of a happy young couple, J.C. and Nannie Meeks. Two-door houses were characteristic of South Carolina textile mill towns; these homes are called duplexes today. (Courtesy of Gloria Hughey.)

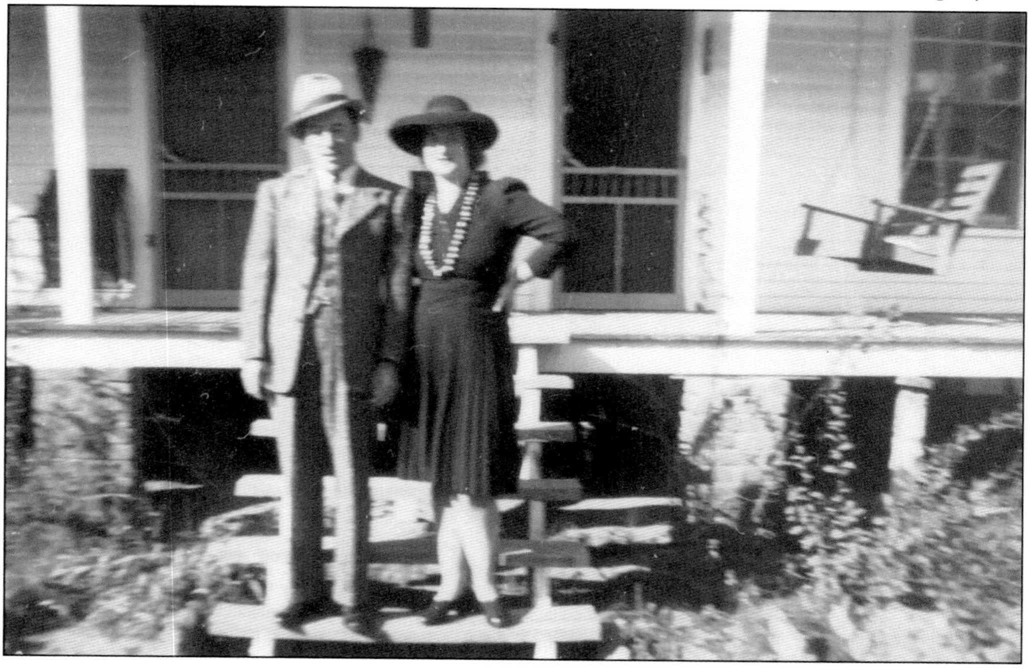

Janie Neves Brashier posed at her family store, Neves Bros. Grocery. The car in the background, wide-legged pants, and the saddle oxfords suggest late 1940s or early 1950s. (Courtesy of Margaret Seel.)

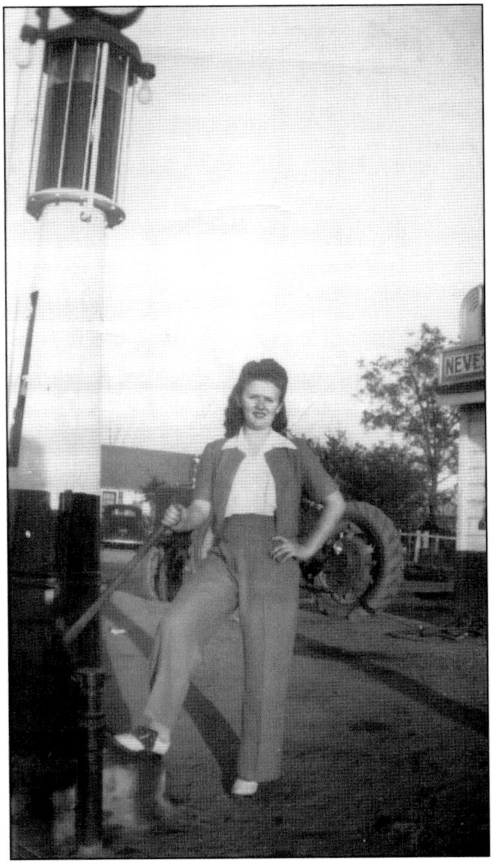

This earlier shot of the Neves family in front of the grocery, which sold Double Cola, includes (in no particular order) Dorothy "Dot" Terry Brashier, Edna Neves Brown, Janie Neves Brashier, and Walton O'Hair Neves. (Courtesy of Margaret Seel.)

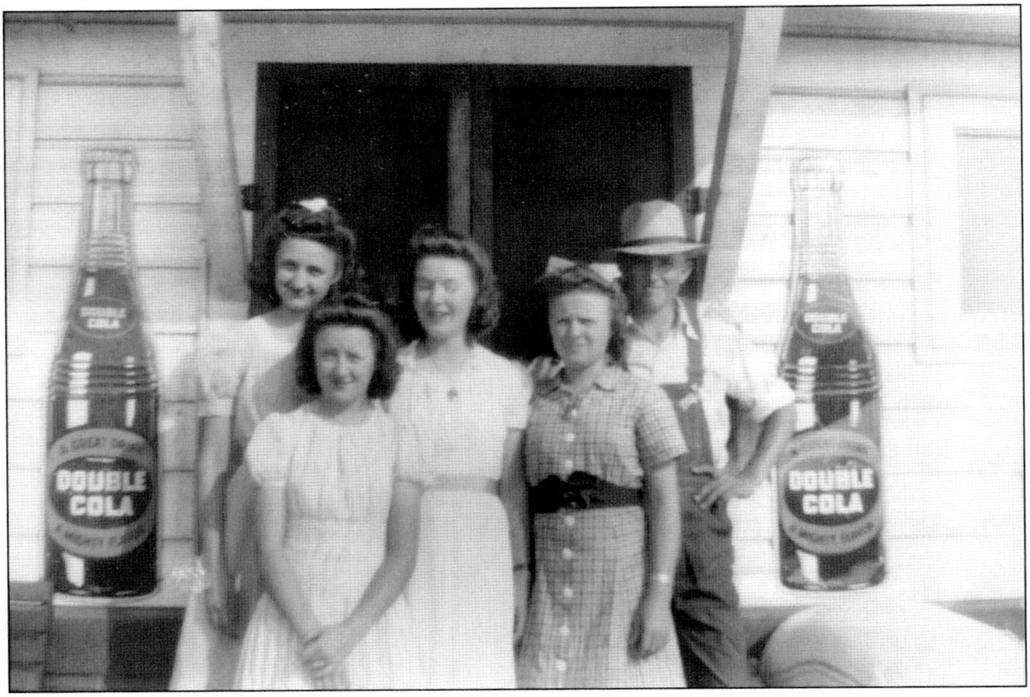

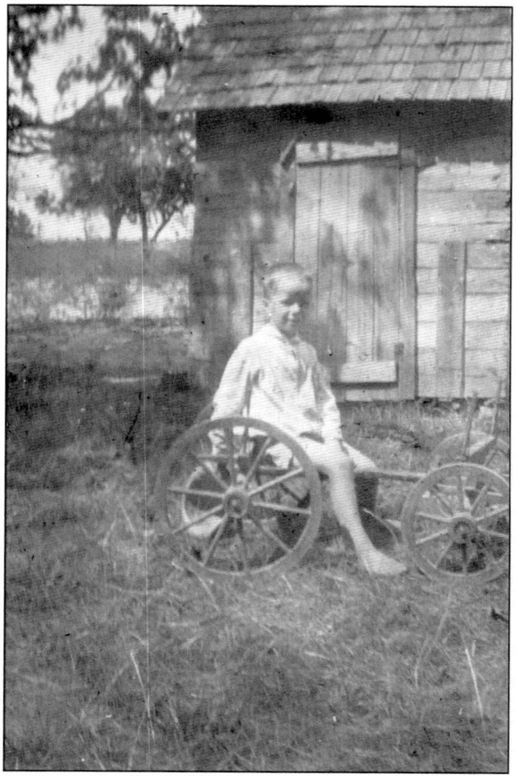

About 1917, Bob Campbell fell ill. During a long recovery, his siblings pulled him around the yard in this cart. In the early years of settling the area and prior to penicillin, many children died young. Two young boys in this family had already been buried. (Courtesy of Anne Peden.)

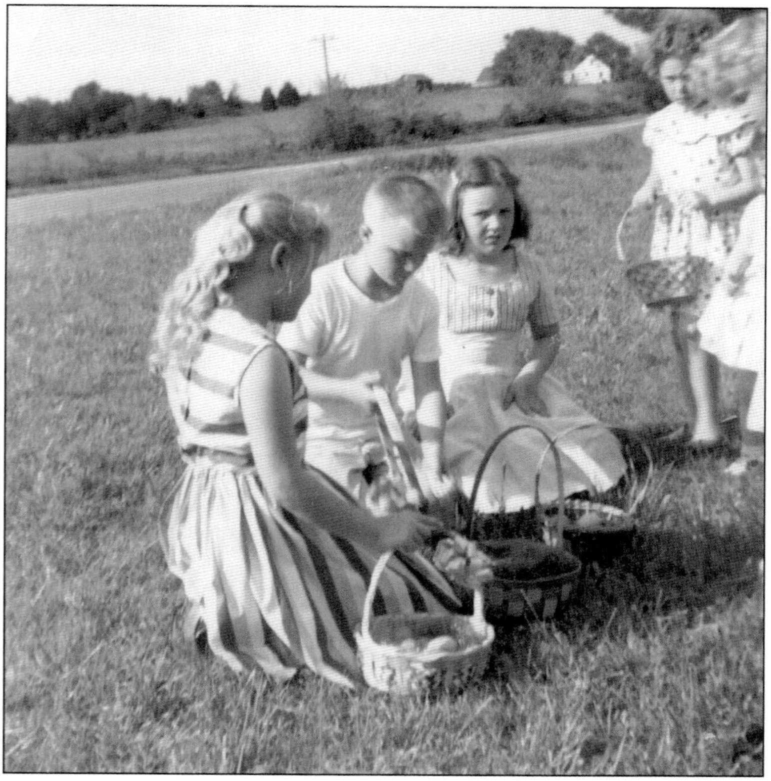

Families raised their children to enjoy the rural life around Fork Shoals. This group of youngsters enjoyed an Easter egg hunt in the 1950s. They are (from left to right) Phillis Brashier Woods, Lonnie Neves, Margaret Ann Neves Seel, Frances Neves Laxton, and Nan Neves Spearman. (Courtesy of Margaret Seel.)

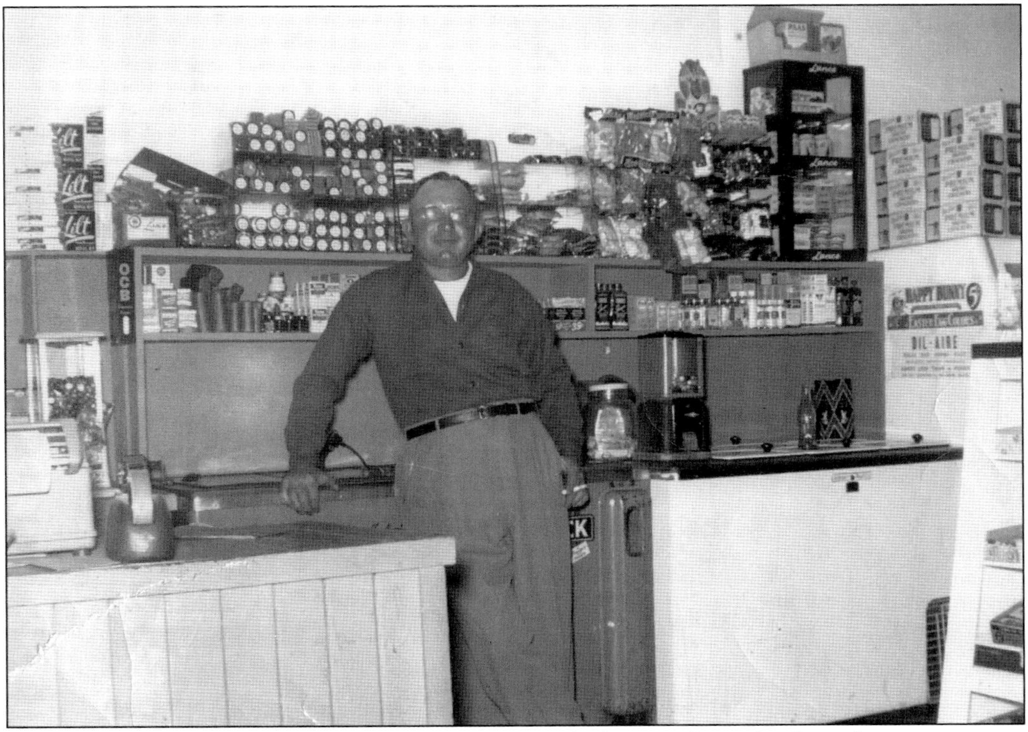

Hilton Smith is standing inside Rhett Hill Store, which was where the fire substation is now. Those gumball machines never change. (Courtesy of Janice Smith Woods.)

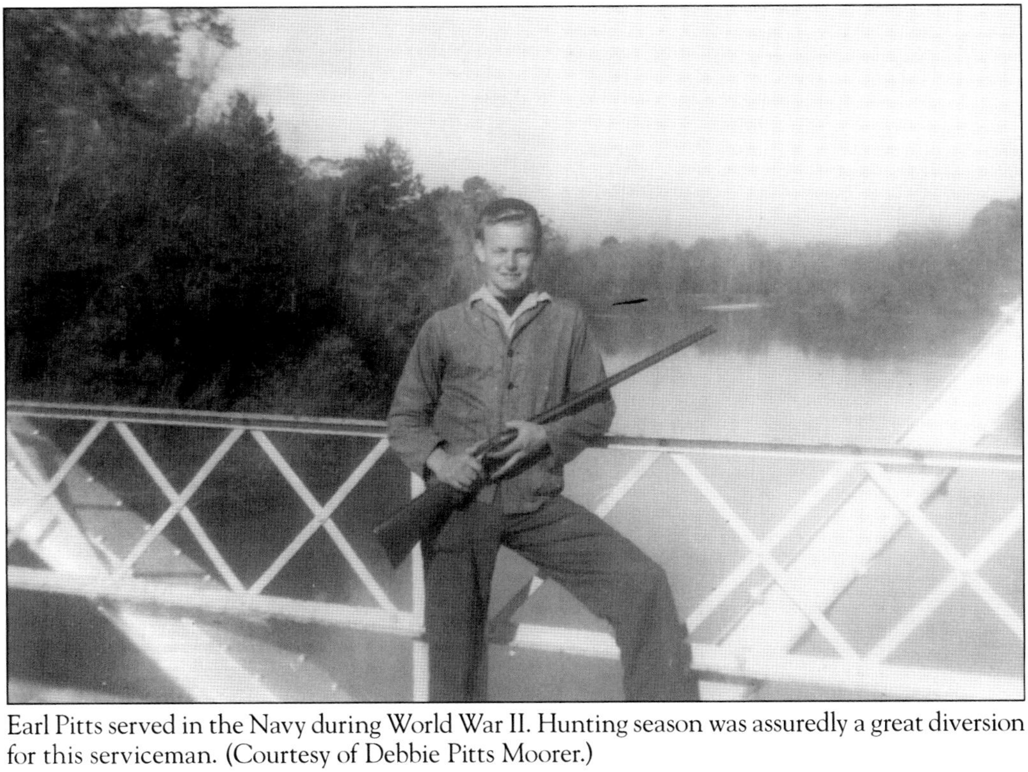

Earl Pitts served in the Navy during World War II. Hunting season was assuredly a great diversion for this serviceman. (Courtesy of Debbie Pitts Moorer.)

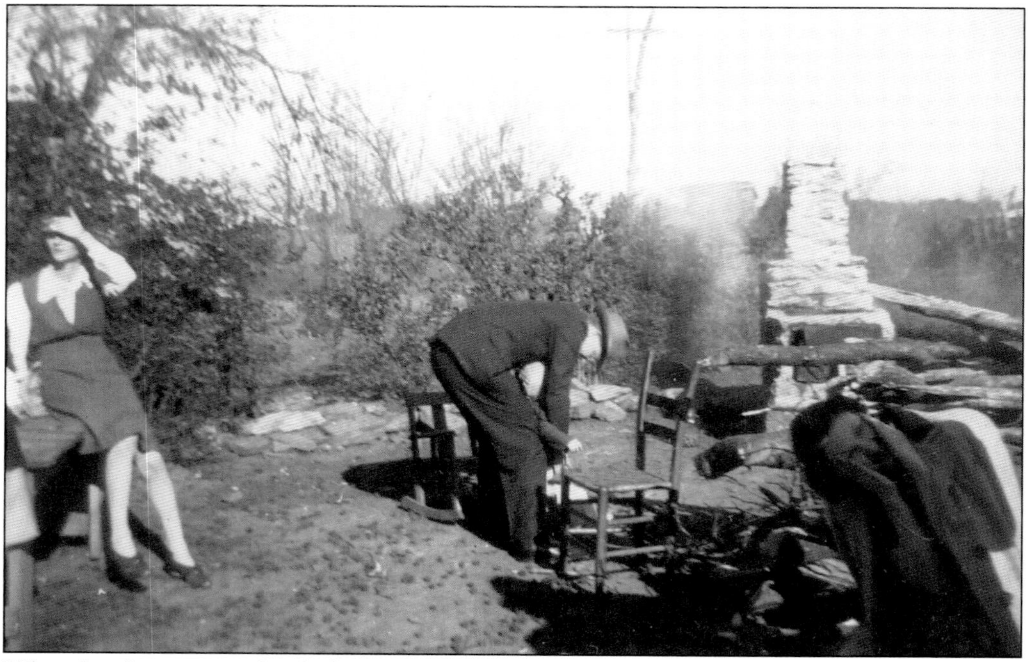

When families got together before air-conditioning became prevalent, they usually migrated out of doors into the vast yards and fields in rural areas. Taken about 1948, this photograph is of such a gathering at the W.E. Campbell home place near Sandy Springs Church. Note that families had the best outdoor fireplaces then too. (Courtesy of Anne Peden.)

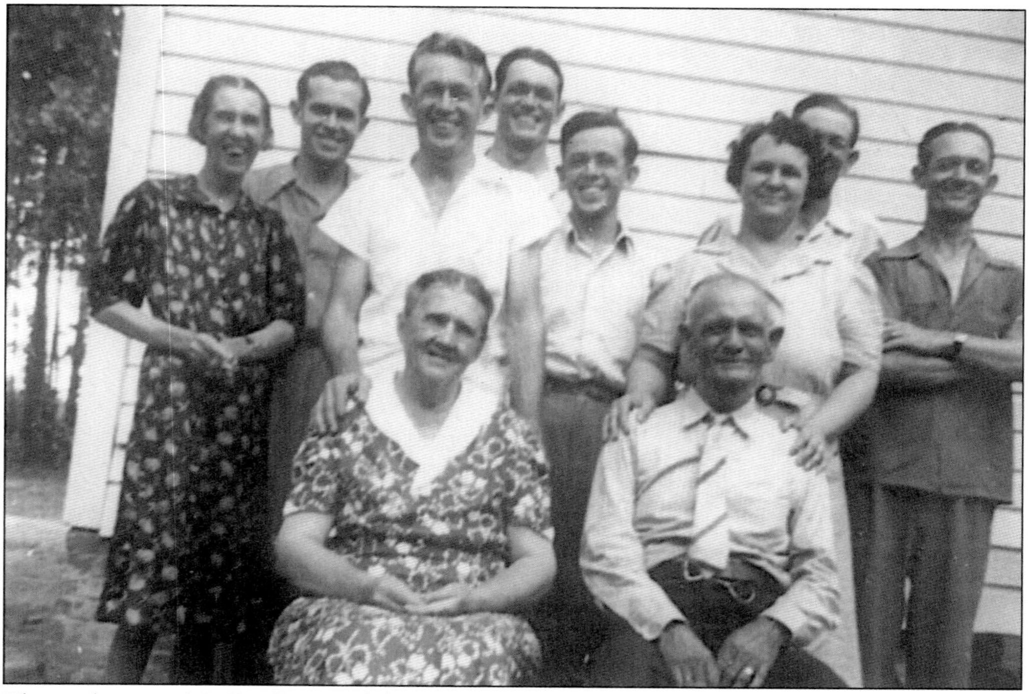

This gathering of Cullen Pitts with his wife and children was taken at his home in Fork Shoals. His son Leroy Pitts (back row, far left) became superintendent of Fork Shoals Mill in 1946. (Courtesy of Debbie Pitts Moorer.)

Hilton Smith is pictured seining in the flood of 1940 at Cedar Falls. This flood was not as destructive as the one in the early 1900s. The freshet of 1908 washed away mills, gins, and stores that had been at the falls for years. After that flood, the businesses were moved up the river to the confluence of Big Creek and the Reedy River. (Courtesy of Janice Smith Woods.)

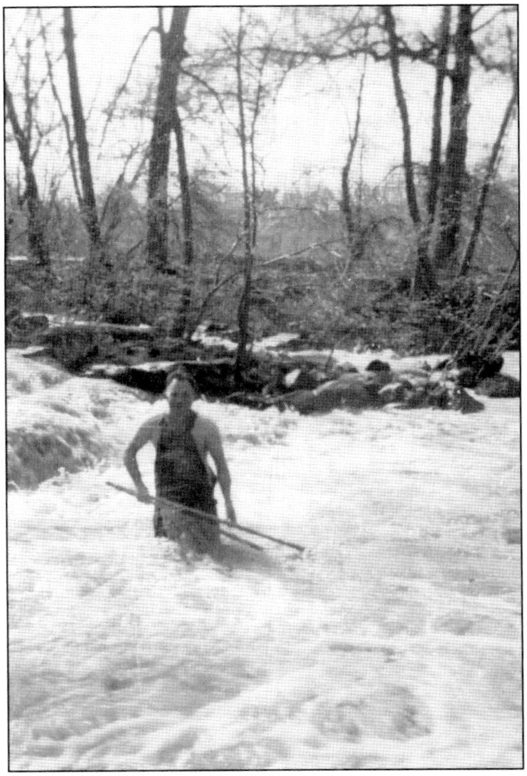

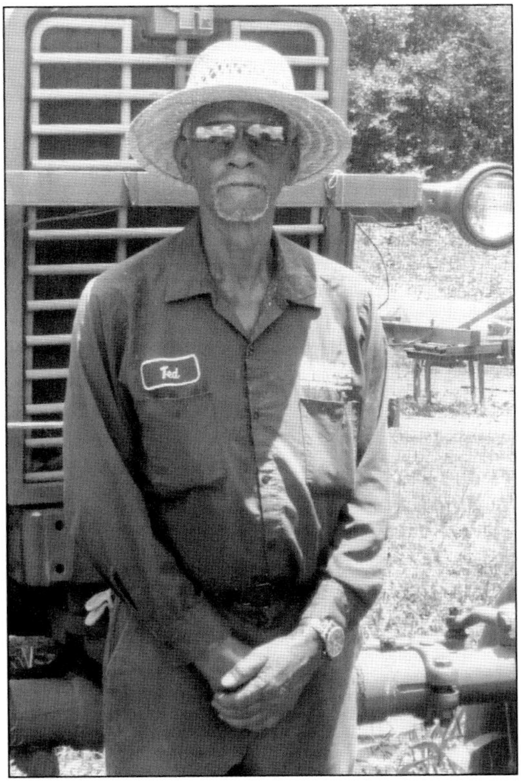

Ted Stewart, son of Jimmie and Lillian Stewart, still lives at their home place and continues to farm the land there. (Courtesy of Virginia Agee Dean.)

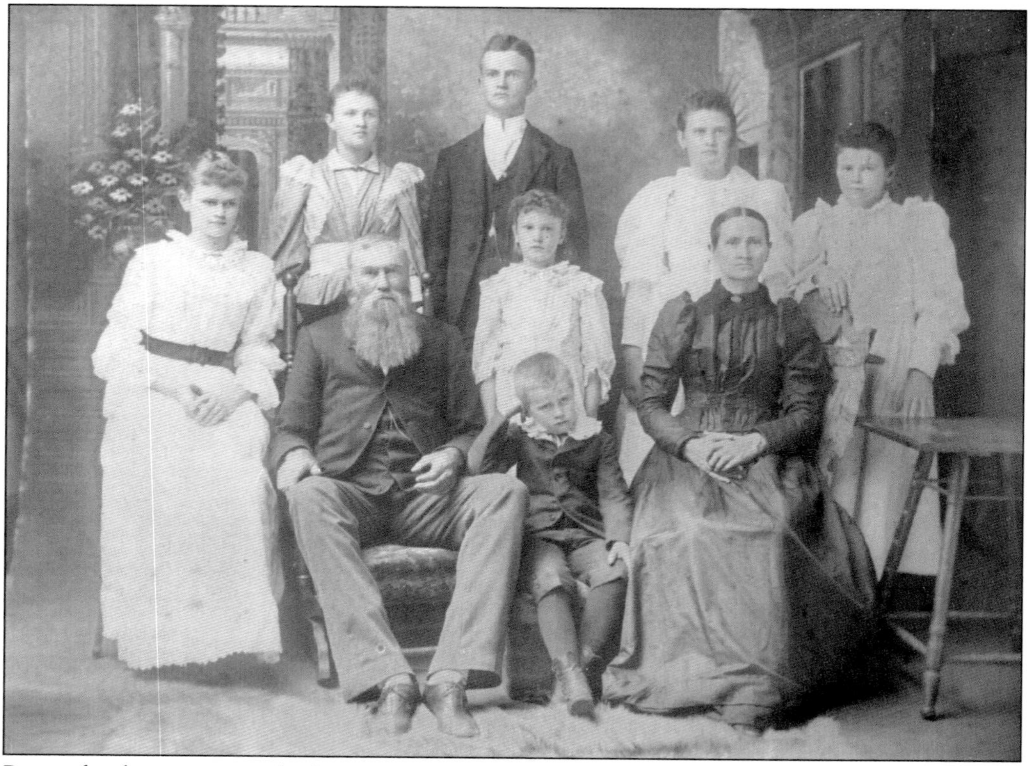

Posing for the camera in the late 1800s is the McKelvey family for whom the main road through Fork Shoals was named. The McKelveys donated the land for building Lickville Church in 1882. Descendants of the family still live on the property. (Courtesy of Lickville Presbyterian Church.)

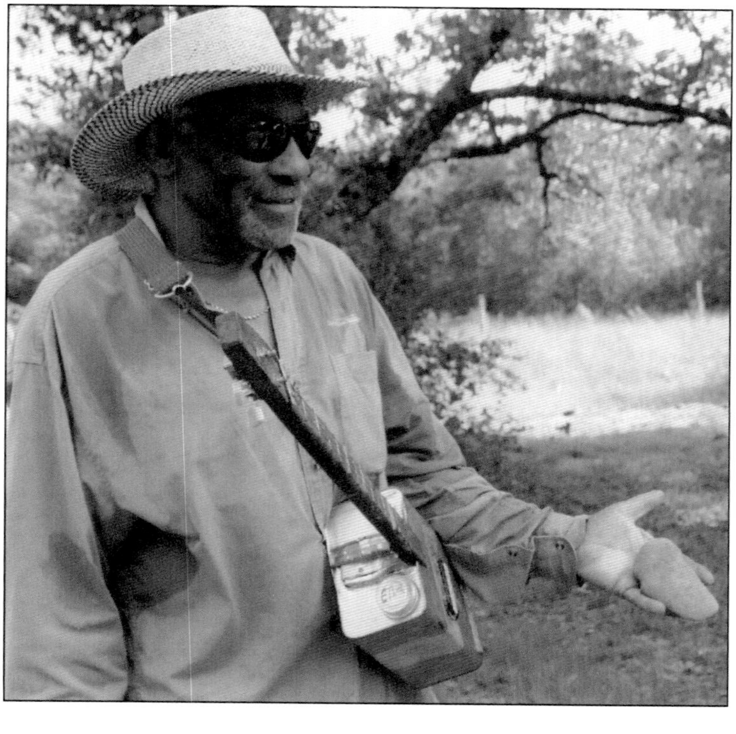

Mac Arnold, blues musician, returned to his family's home place here in 1991. He found this old tomahawk-shaped stone while plowing his fields one day. He and his band support music education in schools and have helped raise funds for music and arts programs. Another goal of his is to encourage organic farming to provide "organic vegetables for the children in our schools." (Courtesy of Cheryl MacKnight.)

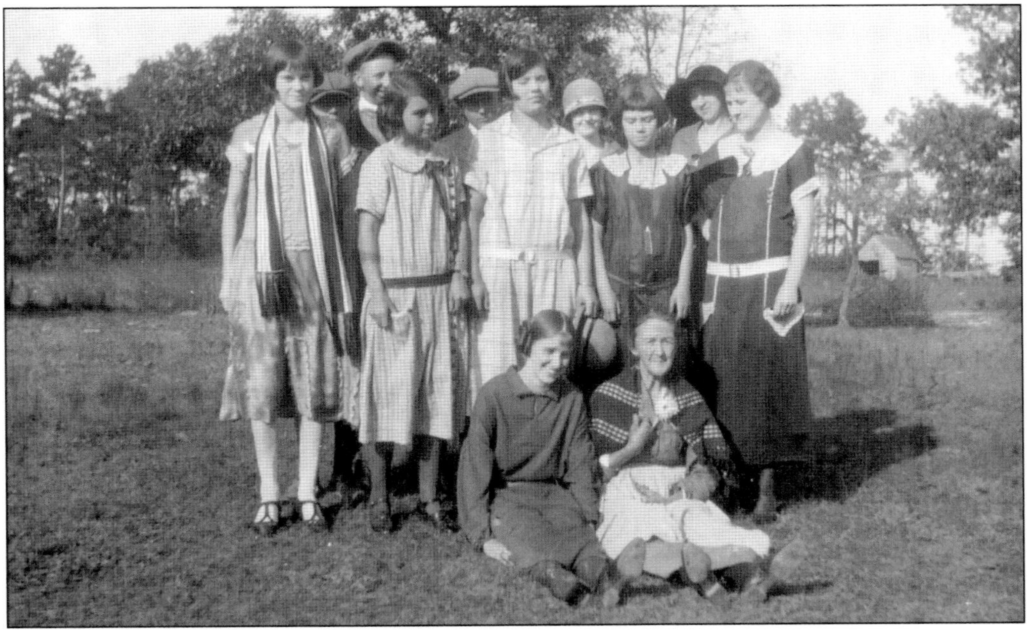

Another gathering brought members of the Jenkins family together for this shot during the 1920s. (Courtesy of Clyde Jenkins.)

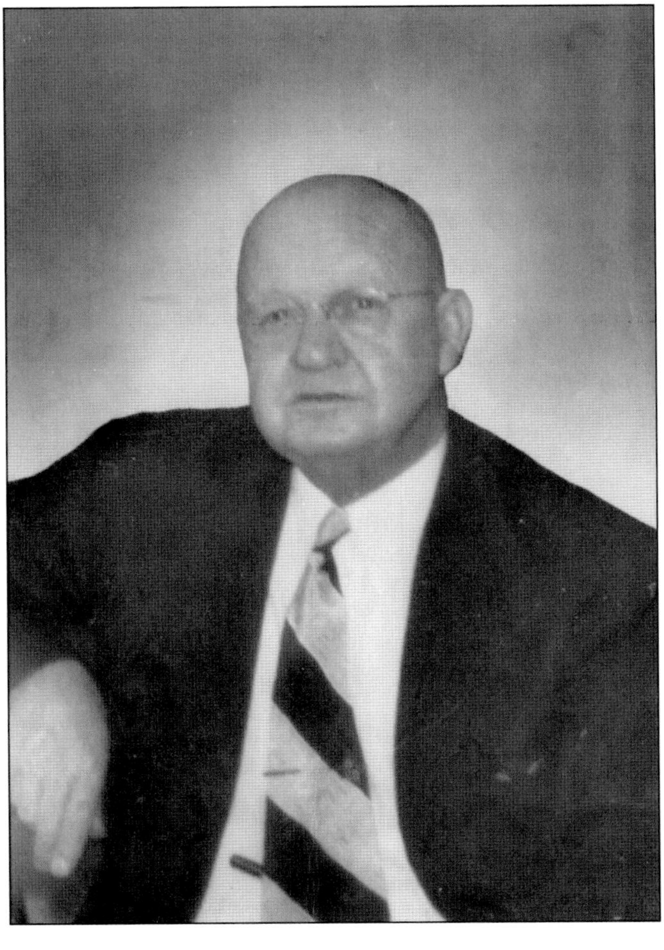

Dr. James Pelham Knight (1889–1951) graduated from Furman University and received his doctor's of medicine degree from Emory University in Atlanta, Georgia. During World War I, Dr. Knight entered the US Medical Corps and served in France. In 1919, he returned home and immediately began serving patients across the southern end of Greenville County. He was at the beck and call of his patients day and night until his sudden death in 1951. Dr. Knight proved himself to be a true servant of mankind. (Courtesy of Frances Knight Horton.)

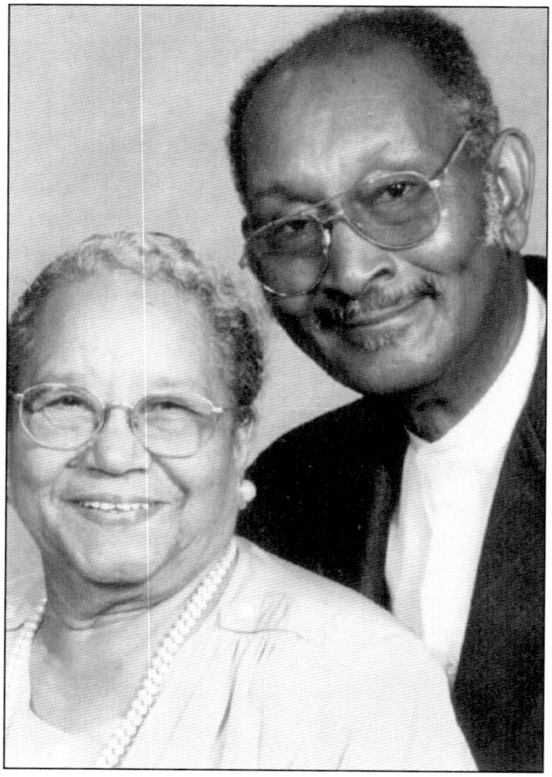

William and Dorothy Thompson Humbert were lifelong members of New Forksville Baptist Church near Princeton. They both served the church in various capacities including deacon and deaconess and Sunday school teachers. Dorothy also presided over the Women's Auxiliary of the Tumbling Shoals Baptist Association. (Courtesy of Truman Humbert.)

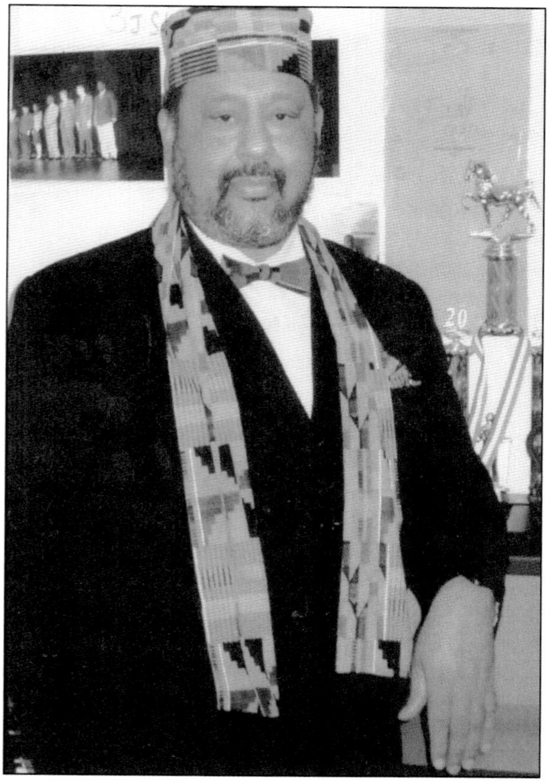

Truman Pelham Humbert was born in 1948 and named after then-president Harry Truman and the doctor who delivered him, Dr. Pelham Knight. He has served in southern Greenville County for many years. He taught US history and psychology at Woodmont High School for 20 years and was nominated as a top-10 finalist for the District Teacher of the Year in 1983. (Courtesy of Truman Humbert.)

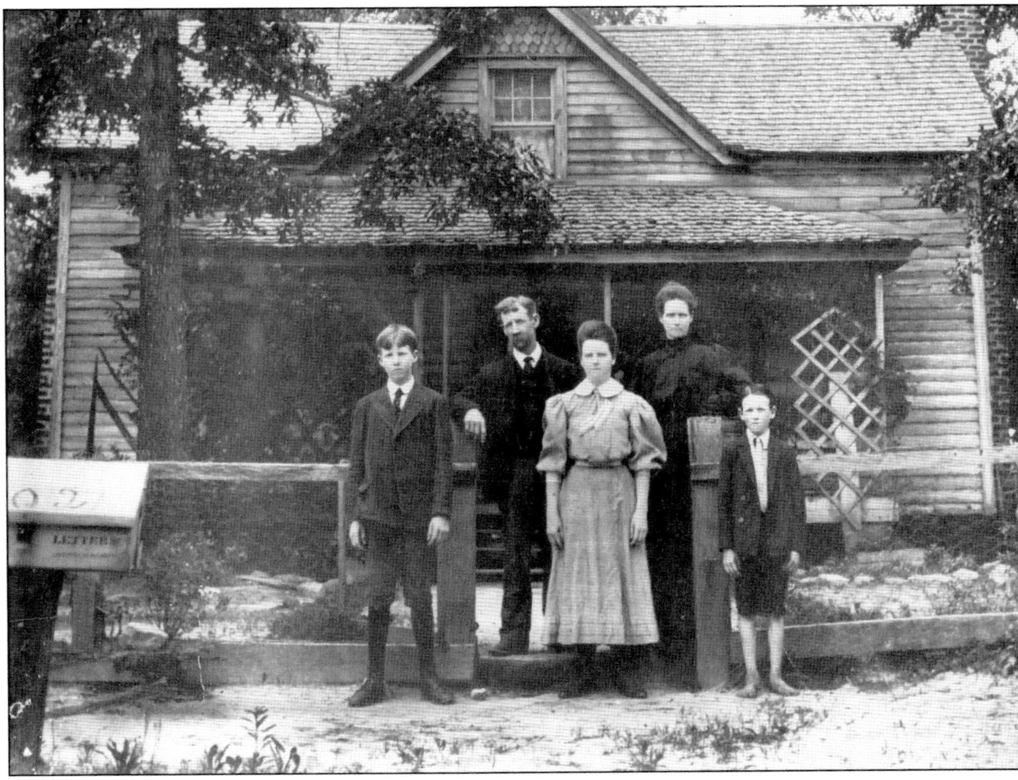

In 1896, Abner Sims purchased this house, built about 1840 by Josiah Chandler. It is still owned by his descendants. Photographed in 1908 were Abner Sims, merchant and farmer, and his wife, Emma Sullivan Sims. Their children were (from left to right) Charles Furman Sims, who became a leader of the Southeastern Baptist Convention; Ollie Sims (Phillips); and W.E. Sims, who became a teacher and superintendent of schools in the area. (Courtesy of Cheryl MacKnight.)

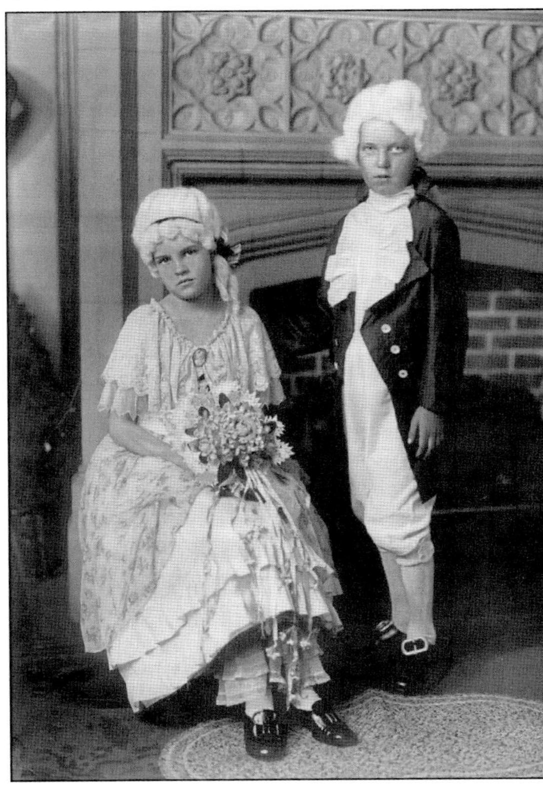

Laodicea "Dicey" Langston was a heroine of the American Revolutionary War. Born in Laurens District in 1759, she, her two brothers, and father fought for the country's independence. Here, Edna Knight Agee, a direct descendant of Dicey, and her cousin Charles Furman Verdin participated in a ceremony in 1933, when a marker was erected at the site of Dicey Langston's home on Tigerville Road. (Courtesy of Virginia Agee Dean.)

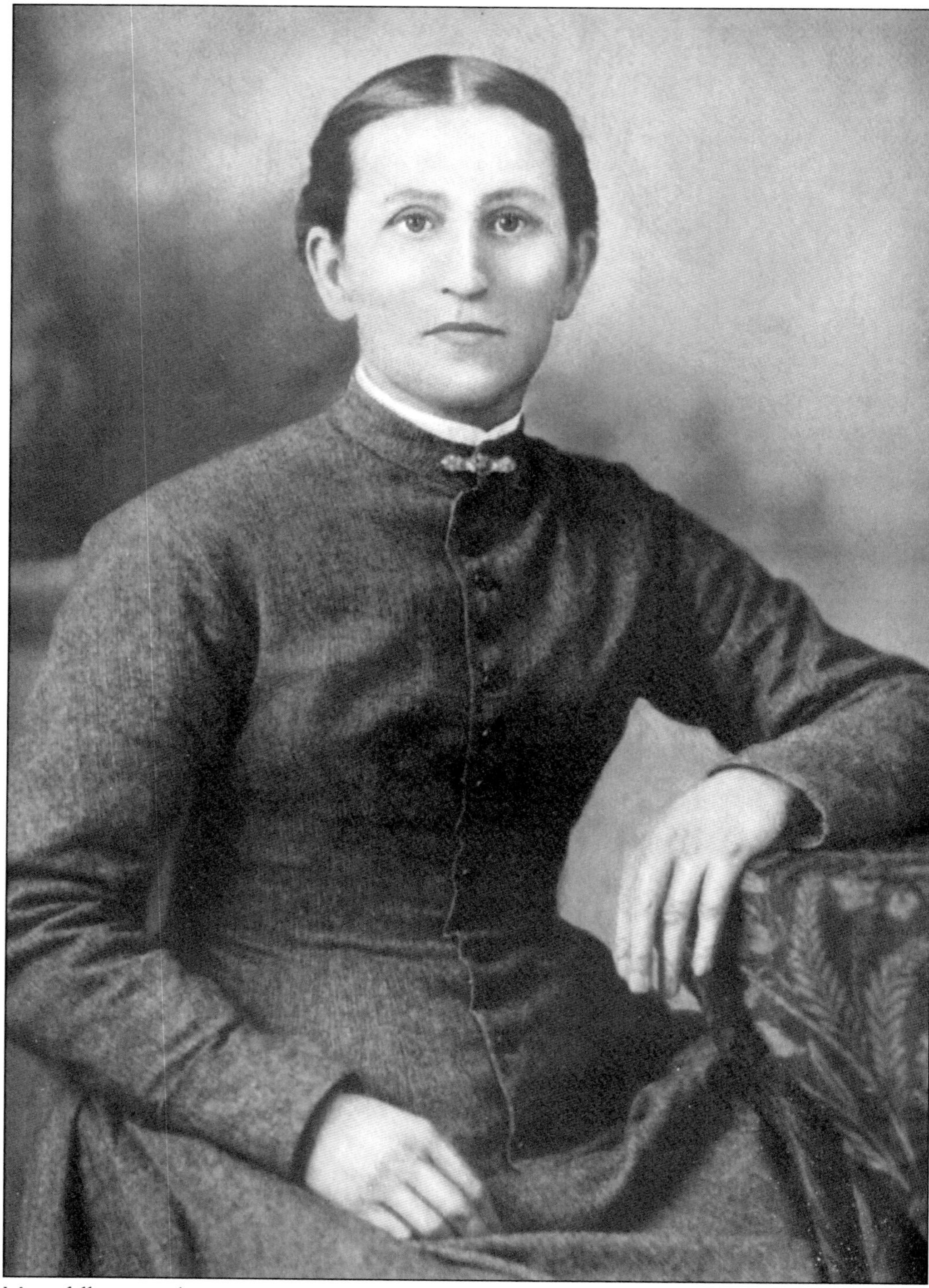

Many folks in southern Greenville County are indebted to Ellen Charles Woodside. After coming to the Augusta Road area from Fairview, she encouraged the establishment of Lickville Presbyterian Church in 1882. Later, the family donated land for the building of a school at Woodville, which was named for her. Ellen Woodside Elementary School is still on the original site. (Courtesy of Lickville Presbyterian Church.)

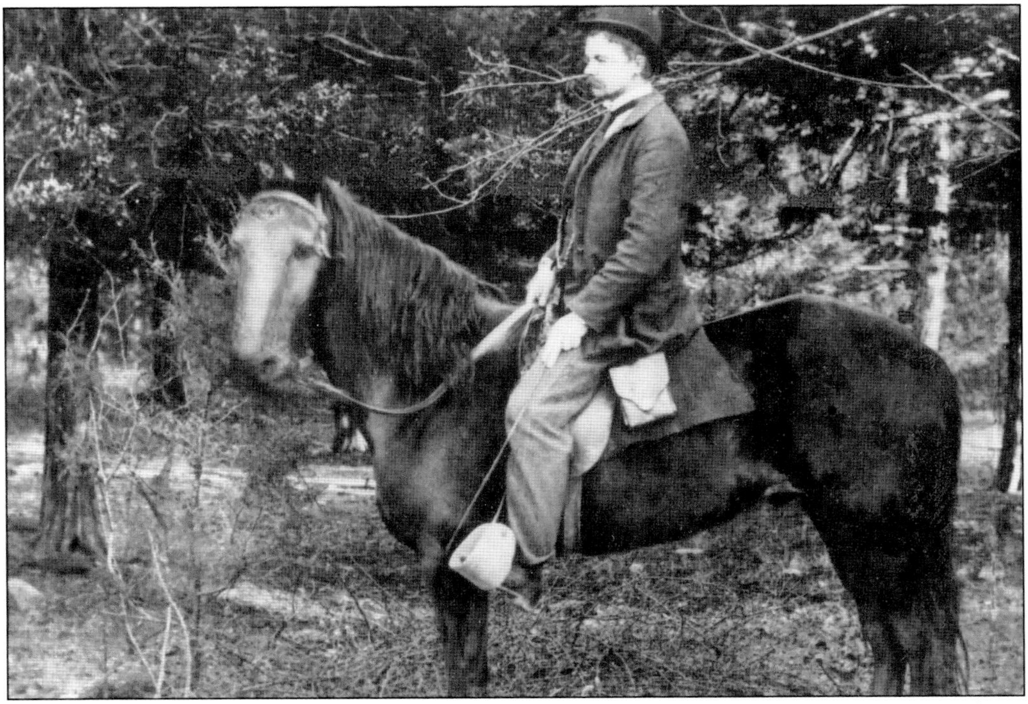

With his medical bag attached to the saddle, Dr. W.A. Ross attended to patients across southern Greenville County during the latter part of the 1800s until his death. His daughter Kitty still lives in their home in Fork Shoals near the elementary school. She was a teacher and librarian there for many years. (Courtesy of Kitty Ross.)

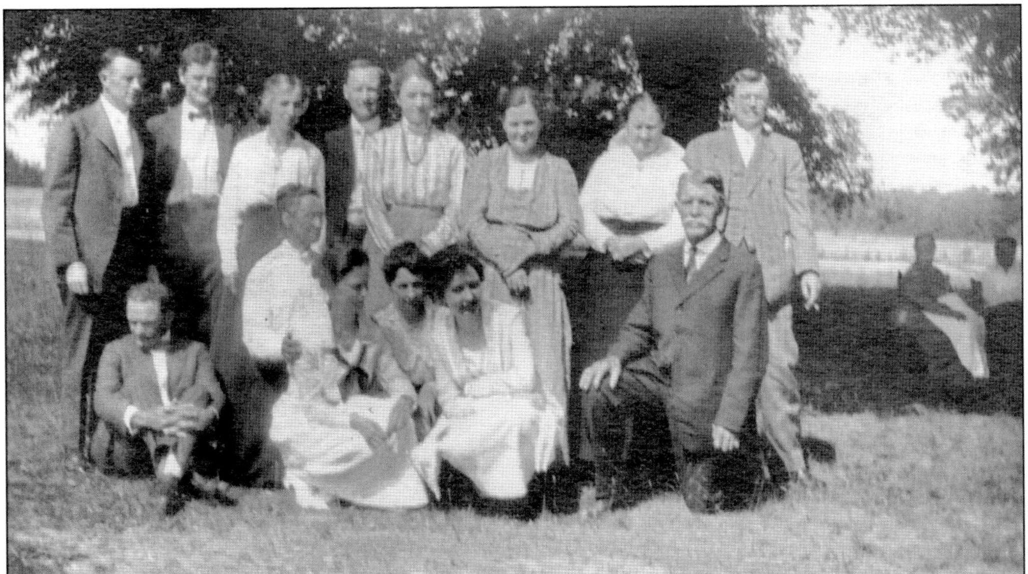

Dr. W.A. Ross's wife, Mary Ellen, was a member of the Scott family, found along the Augusta Road below Lickville Presbyterian Church. This family gathering includes Dr. Ross, far right standing, with other members of the family including Ellie Ross, Edgar and Nancy Babb, Fred and Ida Cox, Tempie McKittrick, Sally and Frank Davenport, and Allen, Robert, Lelia, Kate, and Sue Scott. (Courtesy of Helen Babb Campbell.)

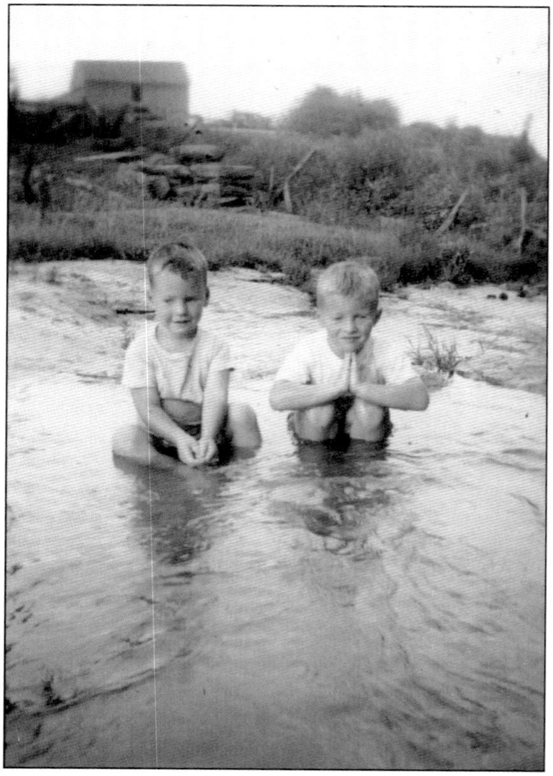

Visiting the creek was an important pastime for children during the summer months. Brothers David and Ralph Richardson are enjoying cooling off in Baker Creek in the 1940s. This creek also has a "wash hole," where several generations of nearby families learned to swim along with the cows from the pasture. (Courtesy of Anne Peden.)

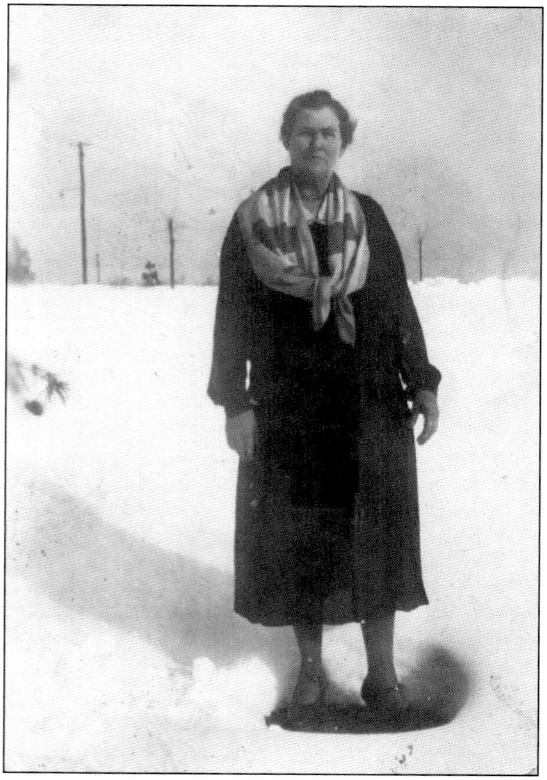

Although summers are warm, with many days above 90 degrees and some above 100, winters can occasionally bring significant snowfall to southern Greenville. Snow was well above Fannie Campbell's ankles here during the 1930s. (Courtesy of Anne Peden.)

Harley Moon's son Leonard mounts his pony and is excited to have the opportunity for a fun weekend ride. Moon and his wife, Elizabeth, and four children, Leonard, Martha, Jane and Sandra, lived on property behind their establishment, the Grill, in Ware Place. They had a lively life as they mingled with those who patronized the café. Other entertainment included a skating rink and go-cart rides. (Courtesy of George and Elaine Moon.)

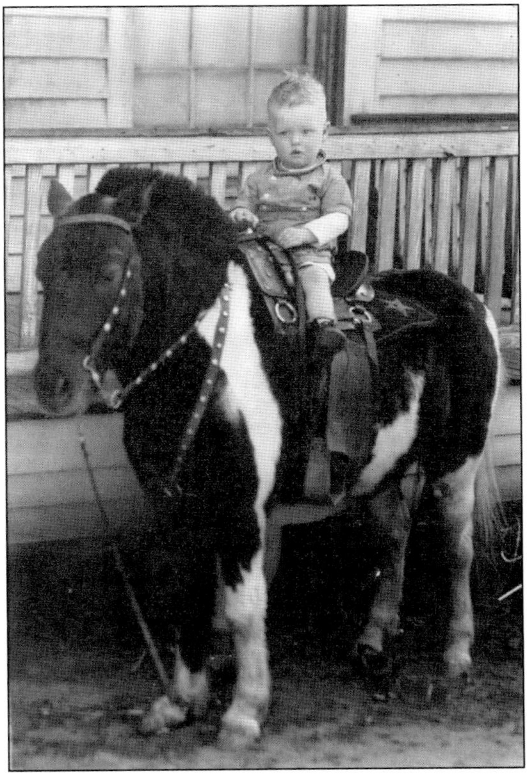

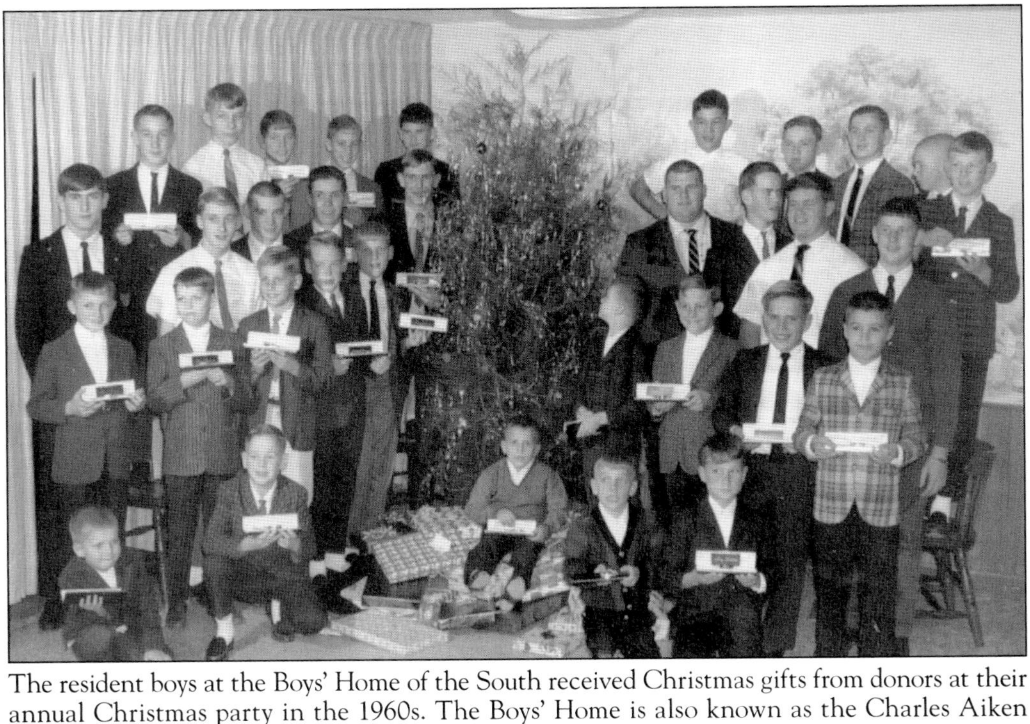

The resident boys at the Boys' Home of the South received Christmas gifts from donors at their annual Christmas party in the 1960s. The Boys' Home is also known as the Charles Aiken Academy in respect for its founder. It has been a safe haven for boys and young men for more than 55 years. (Courtesy of Debbie Pitts Moorer.)

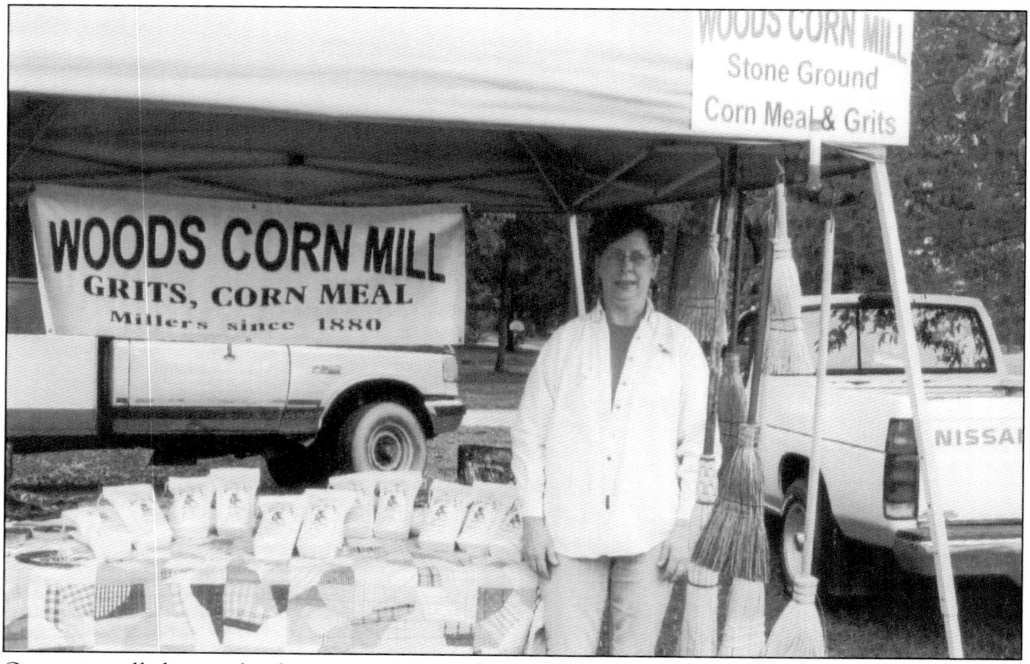

One gristmill that took advantage of the early engines was Woods Corn Mill. Today, the family carries on the tradition of grinding corn into grits and cornmeal. However, the power source is electricity instead of gas engines. Those engines were the source of fires that destroyed two of the Woods Corn Mill locations. (Courtesy of Anne Peden.)

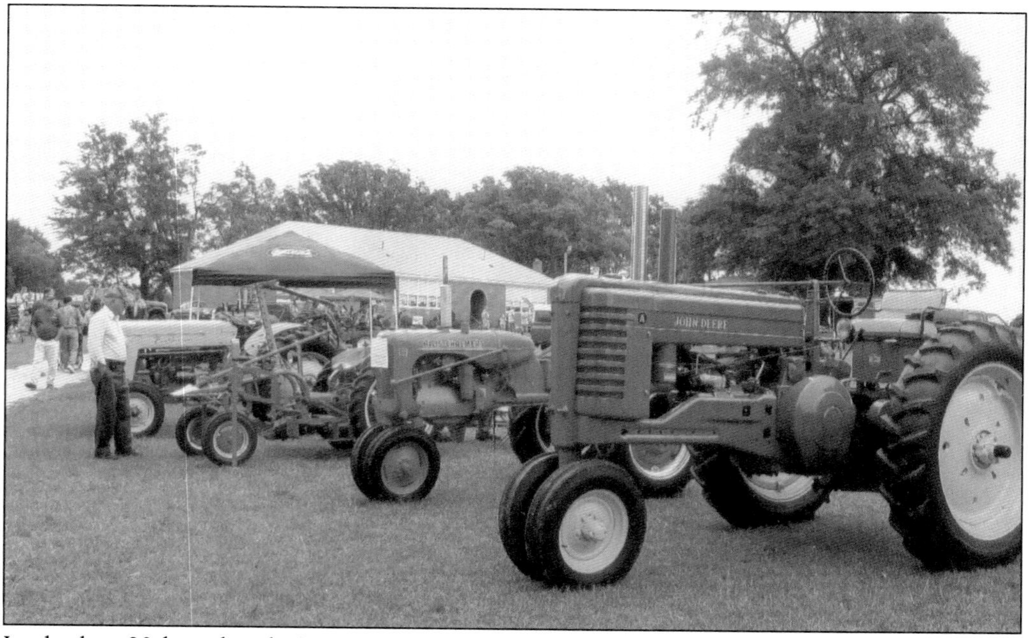

In the late 20th and early 21st centuries, the Fork Shoals community recognized the need to document and celebrate its rich rural heritage. The annual tractor show displays excellent examples of the machines that were typical of the small farms of upstate South Carolina. These pristine John Deere, Allis-Chalmers, and Ford restorations were the tractors found on many of the farms nearby. (Courtesy of Anne Peden.)

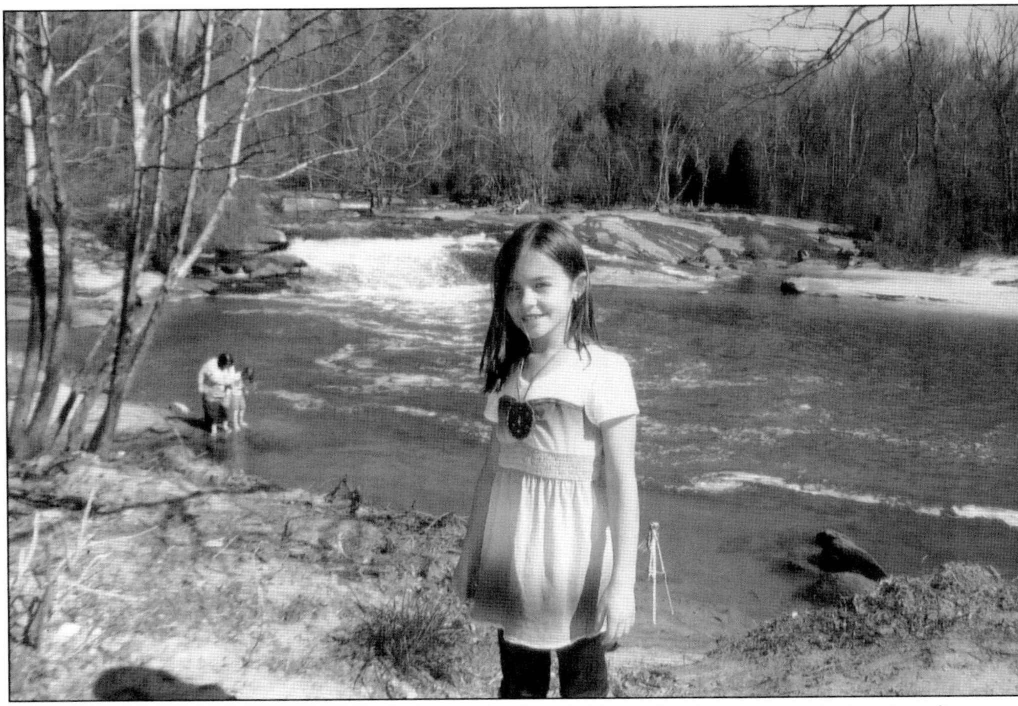

Established in 2011, Cedar Falls Park along the Reedy River has become a destination for many residents as well as visitors to Fork Shoals. Kalyn Cothran enjoys coming to the park often. She and her family live on CaCiba Farms near Fork Shoals Elementary School. (Courtesy of Tuffy Atkins.)

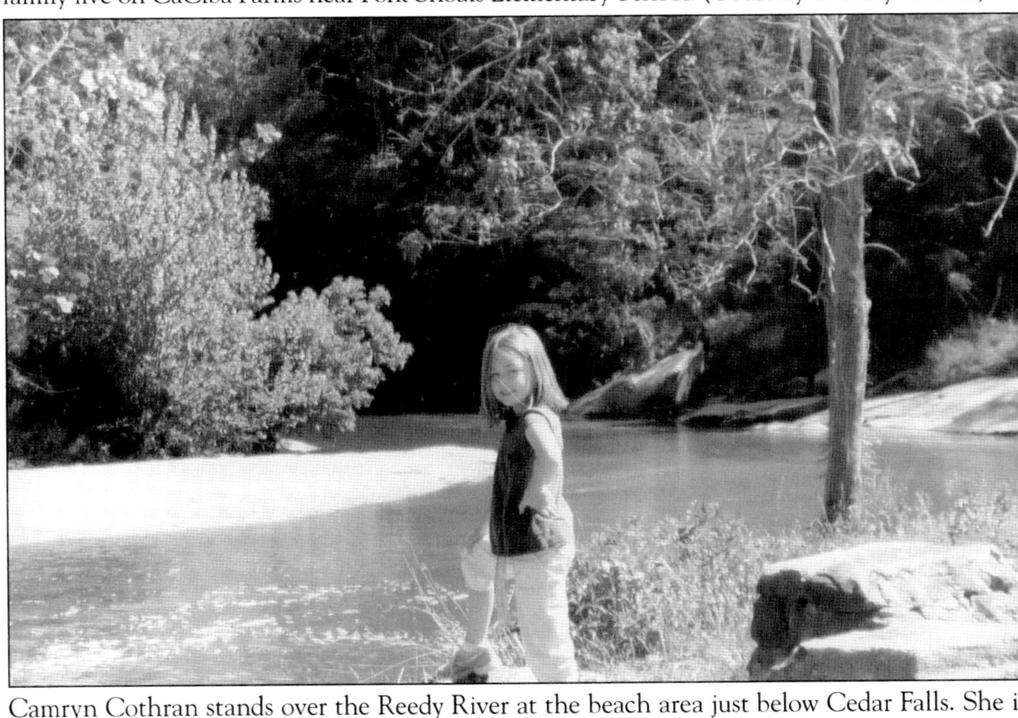

Camryn Cothran stands over the Reedy River at the beach area just below Cedar Falls. She is looking over her shoulder watching for the future generations who will live and thrive in Fork Shoals community. (Courtesy of Tuffy Atkins.)

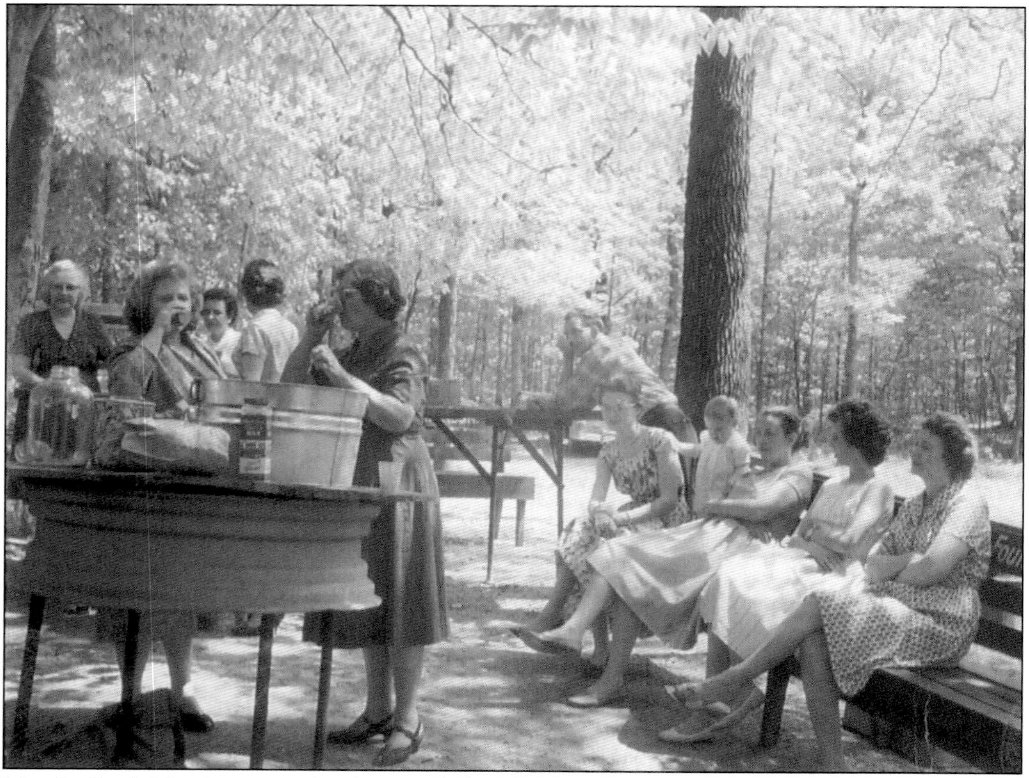

Maude Cook Henderson took a lemonade break at the Fairview Stock Show Fairgrounds with other ladies. Use of a tractor tire rim as a table was a resourceful and clever way to serve the refreshments. People learned to improvise in rural areas, and took pleasure in making do with what was at hand. (Courtesy of Pam Phillips.)

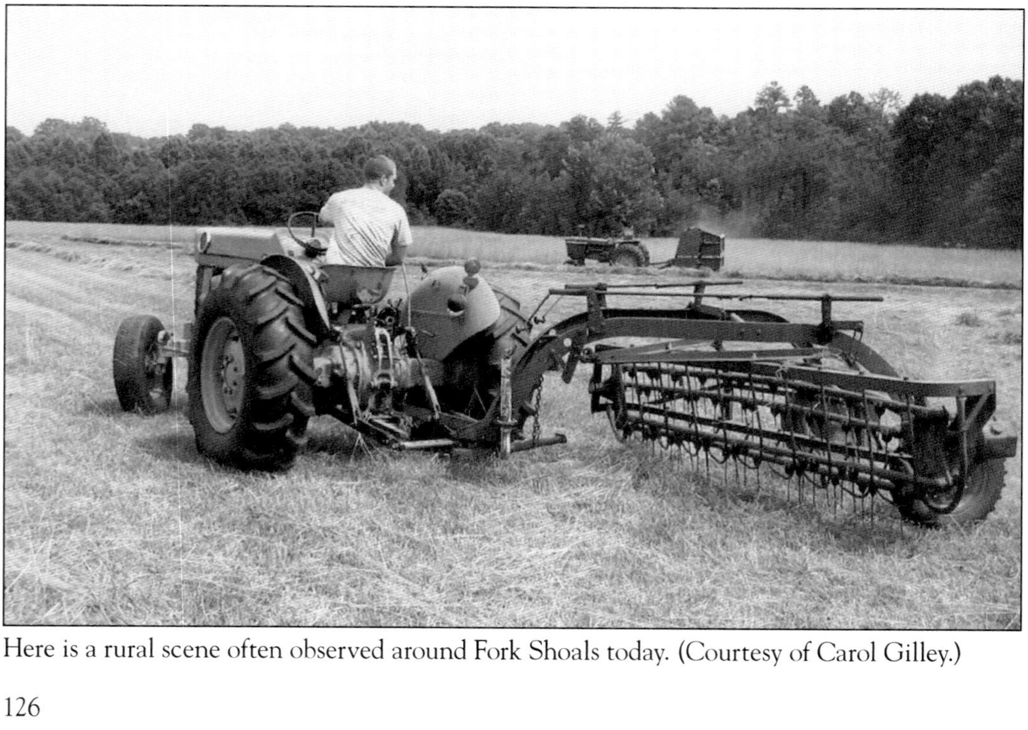

Here is a rural scene often observed around Fork Shoals today. (Courtesy of Carol Gilley.)

About the Organization

The historical society began as the Greenville County Recreation Department worked toward building Cedar Falls Park. Paul Ellis of the recreation district asked for help creating historical signage for the Cedar Falls area in the park. Over that first year, many groups of residents were part of that work as a search for historical photographs began.

The society has worked during its second year to collect more pictures that are significant to the history of the area. This book is the product of that work, and the many residents who have contributed are considered a part of the society.

Our next goal is to catalogue these photographs and documents and to move toward a digital presence in the community. Also, we plan to expand our outreach to enlarge our society with interested individuals.

Please contact one of the committee members if you are interested in working with us in any way. Our mailing address is P.O. Box 442, Pelzer, SC, 29669, or call Cheryl Mac Knight at 864-640-2596. We have members who contribute a wide variety of skills including, collecting, archiving, working at the park, cataloguing birds and other wildlife, and working on digital history.

Join in our work. Do it for Fork Shoals and southern Greenville County. Do it for posterity.

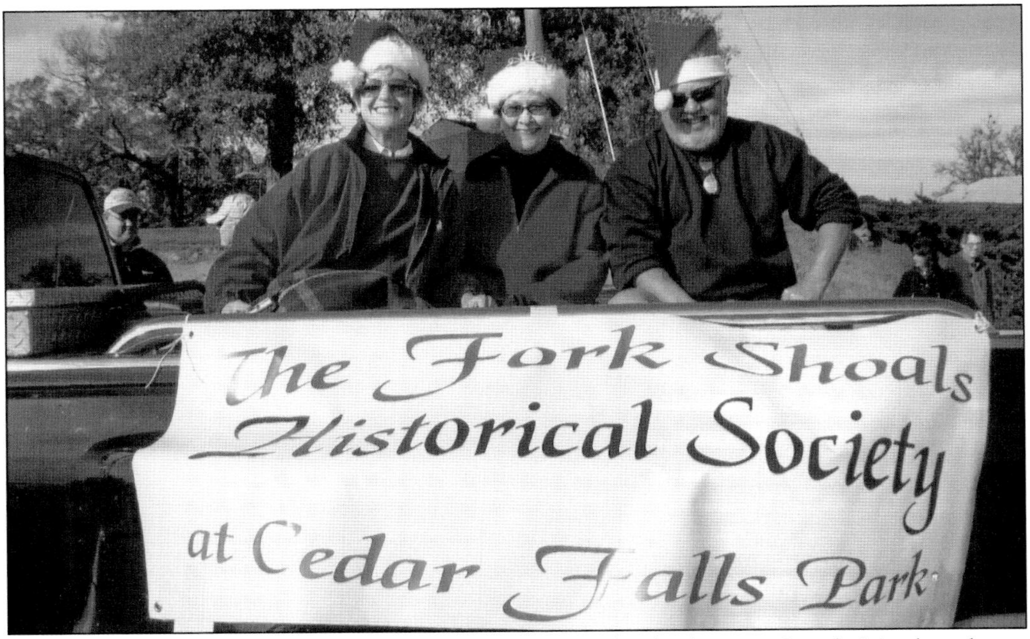

The Fork Shoals Historical Society participates in the Fork Shoals Christmas Parade. Members shown here are, from left to right, Carol Gilley, Tuffy Atkins, and Jim Scott. (Courtesy of Carol Gilley.)